PERSPECTIVES ON SOLID STATE NMR IN BIOLOGY

FOCUS ON STRUCTURAL BIOLOGY

Volume 1

Series Editor
ROB KAPTEIN
Bijvoet Center for Biomolecular Research,
Utrecht University, The Netherlands

Perspectives on Solid State NMR in Biology

Edited by

S.R. KIIHNE and H.J.M. DE GROOT
Leiden University, Leiden, The Netherlands

KLUWER ACADEMIC PUBLISHERS
DORDRECHT / BOSTON / LONDON

A C.I.P. Catalogue record for this book is available from the Library of Congress.

ISBN 0-7923-7102-X

Published by Kluwer Academic Publishers,
P.O. Box 17, 3300 AA Dordrecht, The Netherlands.

Sold and distributed in North, Central and South America
by Kluwer Academic Publishers,
101 Philip Drive, Norwell, MA 02061, U.S.A.

In all other countries, sold and distributed
by Kluwer Academic Publishers,
P.O. Box 322, 3300 AH Dordrecht, The Netherlands.

Printed on acid-free paper

Printed in the Netherlands.

TABLE OF CONTENTS

PREFACE

Most living organisms, including ourselves, are made up of billions of cells separated by membranes containing proteins, the molecular machinery of cells. According to the first bioinformatics estimates from human genome analyses, 30% of all proteins are membrane proteins. Virtually every life process proceeds sooner or later via membrane proteins. Yet, very little is known about how they look and how they work due to a lack of instruments and methods. With thousands of membrane receptor targets awaiting analysis in pharmaceutical companies world wide, the importance of progress in this area is clear.

Solid state NMR is rapidly emerging as a universally applicable method for the characterization of ordered structures that cannot be studied with solution methods or diffraction techniques. Over the past two decades, a number of research groups have worked intensively on developing solid state NMR methods. This work has often been an uphill battle, not only against the usual institutional and natural impediments, but also often against the common opinion of the broader research community. Now the battle is changing, and solid state NMR is becoming an invaluable, widely used technique. There are several reasons why this transition is taking place. First, following explorations of tensor interactions at a fundamental level, a versatile toolbox of basic pulse technologies has been developed. These technologies now act as building blocks in elaborate pulse sequences, much like those currently used in solution state NMR. Second, the NMR hardware has improved considerably with respect to signal to noise, phase stability, high power performance, fast magic angle spinning, etc. These improvements have brought previously unthinkable experiments within the realm of the doable and, in fact, made them commonplace. Third, chemistry and molecular biology now provide sufficient amounts of biological material, often labeled with stable NMR isotopes, to allow specific structural and biological questions to be addressed. Fourth, in the area of biological research, this has allowed a small number of successful demonstrations using solid state NMR technology to answer questions of genuine biological interest, mainly in structural and structure-function studies of membrane proteins, nucleic acids and smaller peptides. Finally, biological understanding of subcellular structures has progressed, revealing important areas of inquiry that require analytical techniques which are applicable to ordered systems without translational symmetry. These areas include drug-receptor complexes involving membrane proteins, low-complexity protein sequences that can form potentially dangerous aggregates *in vivo*, and suprastructures formed by polymers or cellular membranes, among others. In this book, an image of several of these ongoing developments is presented, and together they provide a perspective on the current state of the art of the solids NMR in biology.

The book is a proceedings from the first "Future of Solid State NMR in Biology" conference held at the Gorlaeus Laboratories of Leiden University, Leiden, The Netherlands on October 4-8, 2000. The conference, in part, marked the official opening of the European prototype facility for ultra-high-field Magic Angle Spinning (MAS) NMR at Leiden. The facility is the result of an EU demonstration project, funded by the EC's Biotech program, involving research groups from across Europe. These researchers worked in collaboration with Bruker AG to develop an ultra-high field, wide bore NMR spectrometer and to demonstrate its worth. The spectrometer had been in operation for nearly a year at the time of the conference, and thus, many of the improvements associated with the new technology were first presented there. It is now clear that the implementation of ultra high field magnet technology has provided a significant step forward. Higher fields have led to clear improvements in a broad range of experiments, and the image captured in this proceedings reflects these latest advances and shows some of the ways the new technology can be used to further enhance applications of solid state NMR in biological research.

Extended papers by invited speakers at the conference have been peer reviewed, revised, and edited to present a coherent view of some of the recent progress. We have organized these papers into a number of related sections: First we present papers on NMR pulse sequence developments for applications to non-aligned systems under magic angle spinning conditions. This represents the bulk of today's solid state NMR applications in biology. Secondly, reports of improvements in applications to aligned systems are discussed. These systems hold particular promise for studies of membrane embedded systems and membrane-protein interactions. The third section contains a number of papers that, although closely related to solid state NMR developments, are not directly related to the spectrometer. These include computational methods for simulating NMR data and experiments, and the use of *ab-initio* molecular dynamics studies in interpreting and predicting NMR spectral parameters. This section also presents papers on the expression of integral membrane proteins, in particular G-protein coupled receptors, whose structures and mechanisms of function are central to a very broad range of ailments. There is also a paper on MR microscopy, an area that has been greatly enhanced by the development of high field, wide bore magnets. In the final section, we have gathered together recent applications of state of the art solid state NMR techniques to biological systems of considerable interest. While these applications are only a small slice of the wide variety currently studied by solid state NMR methods, they serve to show what can be accomplished and where the challenges for future development remain.

The conference and the proceedings book would never have been possible without the broad support of the participants and the members of the programme and organising committees. The topics that are presented build upon the extensive body of work done by early pioneers in the field of biological solid state NMR. Although only a few of them have contributed chapters, their work is clearly visible in the theoretical approaches, experimental subjects, and NMR implementation presented by the authors herein. It is a pleasure to stand on the shoulders of these giants.

This book is intended for researchers and students interested in broadening their understanding of current solid state NMR methods and techniques. It shows some of the things that can be achieved with higher field magnets in the realm of solid state NMR, and thus provides an up-to-date starting point for developing new projects and applications: a jumping off point for those who want to be part of the future of solid state NMR in biology.

Leiden, April 25, 2001   Suzanne R. Kiihne
                                                                            Huub J.M. de Groot

International Programme Committee:
Prof. dr. M. Baldus, Germany
Dr. B. Bechinger, Germany
Dr. H. Foerster, Germany
Prof. dr. W.J. de Grip, The Netherlands
Prof. dr. H.J.M. de Groot, The Netherlands
Prof. dr. A.R. Holzwarth, Germany
Dr. S.R. Kiihne, The Netherlands
Prof. dr H. Oschkinat, Germany
Prof. dr. R. Poelmann, The Netherlands
Prof. dr. A. Watts, United Kingdom

Local Organizing Committee:
Dr. C.J. Bloys van Treslong
Prof. dr. W.J. de Grip, (Treasurer)
Prof. dr. H.J.M. de Groot, (Vice-chairman)
Dr. F. Heidekamp
Dr. S.R. Kiihne
Prof. dr J. Lugtenburg, (Chairman)
Prof. dr. R. Poelmann
H. E. Schat-Hansen, B.A., (Secretary)

Conference Staff
Marc Beck
Dr. Hugo Zwenk

Speakers at the Conference:

M. Baldus	Max-Planck-Institute for Biophysical Chemistry, Göttingen,		Germany
B. Bechinger	Max-Planck-Institute for Biochemistry, Martinsried		Germany
G.C. Bosman	KUN Faculty of Medical Sciences	Nijmegen	The Netherlands
F. Buda	Vrije Universiteit	Amsterdam	The Netherlands
A. Creemers	Leiden University	Leiden	The Netherlands
T. A. Cross	National High Magnetic Field lab.	Tallahassee, FL	U.S.A.
G. Drobny	University of Washington	Seattle, WA	U.S.A.
T. Egorova	Leiden University	Leiden	The Netherlands
C. Glaubitz	Oxford University	Oxford	U.K.
R.G. Griffin	Mass. Institute of Technology	Cambridge, MA	U.S.A.
A. Haase	Universität Würzburg	Würzburg	Germany
J. Herzfeld	Brandeis University	Waltham, MA	U.S.A.
B. Hogers	Leiden University Medical Center	Leiden	The Netherlands
H. Kiefer	m-phasys GmbH	Tübingen	Germany
A. Killian	University Utrecht	Utrecht,	The Netherlands
M.H. Levitt	Stockholm University	Stockholm	Sweden
K. Lundstrom	F. Hoffman-La Roche Ltd.	Basel	Switzerland
J. Matysik	Leiden University	Leiden	The Netherlands
B.H. Meier	ETH Zentrum	Zurich	Switzerland
N. C. Nielsen	University of Aarhus	Aarhus	Denmark.
S. J. Opella	University of Pennsylvania	Philadelphia, PA	U.S.A.
J. Pauli	Forschungsinstitut Für Molekulare Pharmakologie, Berlin,		Germany
J.H.C. Reiber	Leiden University Medical Center	Leiden	The Netherlands
B.J. van Rossum	Institut Für Molekulare Pharmakologie (FMP), Berlin		Germany
J. Schaefer	Washington University	St. Louis, MO	U.S.A.
A. Schmidt	Israel Institute Of Technology	Haifa,	Israel
A. Sebald	Universität Bayreuth	Bayreuth	Germany
B.R. Smith	University of Michigan	Ann Arbor, MI	U.S.A.
S. O. Smith	SUNY Stony Brook	Stony Brook, NY	U.S.A.
G.J. Turner	University of Miami	Miami, FL	U.S.A.
D. Steensgaard	MPI Für Strahlenchemie	Muelheim a/d Rhur,	Germany
G. Gröbner	University of Umea	Umea	Sweden
A. S. Ulrich	University Jena	Jena	Germany
H. Van As	Wageningen University	Wageningen	The Netherlands
S. Vega	Weizmann Institute	Rehovot	Israel
P. Williamson	ETH-Zentrum	Zurich,	Switzerland

SECTION I:

Advances in MAS Techniques for Non-Aligned Systems

Using symmetry to design pulse sequences in solid-state NMR

Andreas Brinkmann, Marina Carravetta, Xin Zhao, Mattias Edén, Jörn Schmedt auf der Günne, and Malcolm H. Levitt
Physical Chemistry Division, Stockholm University, S-10691 Stockholm, Sweden

Introduction

Modern solid-state NMR employs a range of rf pulse sequences for a variety of tasks. There are *decoupling* sequences which reinforce the averaging effect of the magic-angle rotation, causing different spin species to evolve approximately independently of each other. There are also *recoupling* sequences which undo the averaging effect of the magic-angle rotation, temporarily restoring couplings which are otherwise inactivated by the sample spinning. The success of solid-state NMR in biological research may depend on the development of decoupling and recoupling pulse sequences which are robust with respect to a variety of undesirable spin interactions and experimental imperfections, and which function over a wide range of static magnetic fields and/or spinning frequencies.

With this in mind, our group has developed a general pulse sequence design strategy which is applicable to a wide range of methodological situations. The strategy is based on the use of symmetry arguments to design rotor-synchronized pulse sequences[1-7]. The basic decoupling and recoupling properties of a certain types of pulse sequence may be predicted using the values of just three integers, called symmetry numbers. These results apply even when the spinning frequency is comparable to the nutation frequency of the spins in the rf field. Furthermore, the symmetry theorems may be inverted: It is possible to find the combinations of symmetry numbers that give rise to a set of desirable pulse sequence properties. The symmetry theory provides a theoretical framework which greatly simplifies the task of designing rotor-synchronized pulse sequences. In this short article, we sketch the basic symmetry arguments and illustrate a few of the pulse sequences we have designed and demonstrated. In addition, we show that the symmetry theorems provide insight into the behaviour of a number of

S.R.Kiihne and H.J.M.deGroot (eds.), Perspectives on Solid State NMR in Biology, 3-14.

Table 1. Classification of spin interactions in a homonuclear system. The interactions are: Isotropic chemical shift; J-coupling; chemical shift anisotropy; direct dipole-dipole coupling.

	Space Rank l	Spin Rank λ	Field Rank
Iso-CS	0	1	1
J	0	0	0
CSA	2	1	1
DD	2	2	0

existing pulse sequences. Much of this work is of a preliminary nature and will be described in more detail in forthcoming publications.

Theory

The symmetry theorems are based on the rotation properties of the spin interaction terms. Table 1 summarizes the rotation properties of the most important nuclear spin interactions for homonuclear coupled systems (we have also generalized our results to heteronuclear systems, but this will not be discussed here). Each interaction term behaves as a component of an irreducible spherical tensor with respect to rotations of either the molecules ("space rank", denoted l), the nuclear spin polarization axes ("spin rank", denoted λ), or the static magnetic field ("field rank"). In practice, the "field rank" is unimportant since the direction of the static field cannot normally be changed, for technical reasons. Nevertheless, the field rank is included in Table 1 for completeness. The table shows that the pairs of space-spin ranks are different for all the interactions. The symmetry theorems described below exploit this distinction by synchronized "space rotations" (i.e. bulk rotation of the whole sample) and "spin rotations" (implemented by rf field pulse sequences).

In general a spherical tensor of rank l has $2l+1$ spherical components, with indices $m=-l, -l+1..+l$. Similarly, each spin interaction has a set of components, with space quantum numbers m and spin quantum number μ taking all possible values permitted by the space and spin ranks l and λ. In the presence of a rf field, the homonuclear DD interaction has 25 components (5×5), while the CSA interaction has 15 components (3×5).

The large number of rotational components contributes to the technical difficulty of the general recoupling problem. As an example, consider the task of double-quantum homonuclear recoupling, which is one of the most useful modes of reintroducing the homonuclear dipole-dipole interactions [1-9]. In this experiment, the evolution of the spin system should be rendered sensitive to only the double-quantum DD components with $\mu=-2$ and +2. The chemical shift terms and other DD terms should be suppressed. In addition, it is often desirable that each spin component μ is only associated with a single space component m, since this leads to a favourable orientation-dependence of the recoupling (this property is called γ-encoding [1-3, 9]). In order to create a γ-encoded 2Q recoupling sequence, it is therefore necessary to suppress over 30 undesirable spin interaction terms, retaining only two. Any attempt to do this by searching over the possible combinations of pulse flip angles and rf phases, without the assistance of symmetry, has little hope of success.

C-type sequences. So far, the symmetry theorems apply to two general classes of rotor synchronized pulse sequence. The first class is denoted CN_n^ν, where N, n and ν are the three symmetry numbers [4, 6, 7]. This class of pulse sequence may be constructed as follows: (i) take n sample revolution periods; (ii) divide this into N equal intervals; (iii) fill each interval with a rf pulse sequence C, which is a *cycle*, meaning that the rf fields bring the spins back to their starting states, if all other interactions are ignored; (iv) shift the phase of the qth C element by the angle $2\pi q\nu/N$, where q=0,1..N–1. The symmetry numbers n and ν are called the *space* and *spin winding numbers* [6].

The selection rule for CN_n^ν symmetry may be denoted [4, 6]

$$\overline{H}^{(1)}_{lm\lambda\mu} = 0 \text{ if } mn-\mu\nu \neq NZ \tag{1}$$

where Z is any integer (including zero). The left-hand side of this equation expresses the first-order average Hamiltonian for the term with space rank l, spin rank λ and component indices m and μ. If a term with a certain combination of quantum numbers $lm\lambda\mu$ vanishes, then the pulse sequence is expected to be insensitive to this term, to a good approximation. By choosing values of N, n and ν that cause Eq.(1) to be satisfied for all the *undesirable* terms $lm\lambda\mu$, while ensuring that Eq.(1) is *not* satisfied for all the *desirable* terms, one may focus on pulse sequences which are guaranteed to be good candidates for the task in hand.

Space-spin selection diagrams [6] are useful for visualizing the consequences of Eq.(1). One example is shown for the symmetry $C7_2^1$ in Fig.1. For each interaction, the composition of the term mn-μν is broken into two stages, so as to make clear how the symmetry numbers n and ν interact. The inequality in Eq.(1) is visualized by a barrier with holes separated by N units. Only interactions which "pass through the holes" do not satisfy the inequality in Eq.(1) and are retained in the average Hamiltonian. Interactions which "run into the wall" are suppressed in the average Hamiltonian. The diagram in Fig.1 illustrates that the symmetry $C7_2^1$ suppresses all terms except for the 2Q DD terms $(lm\lambda\mu)=(2122)$ and (2 -1 2 -2), and hence has the basic properties necessary for γ-encoded 2Q recoupling with chemical shift suppression. The successful C7 [1] and POST-C7 [2] sequences exploit this symmetry.

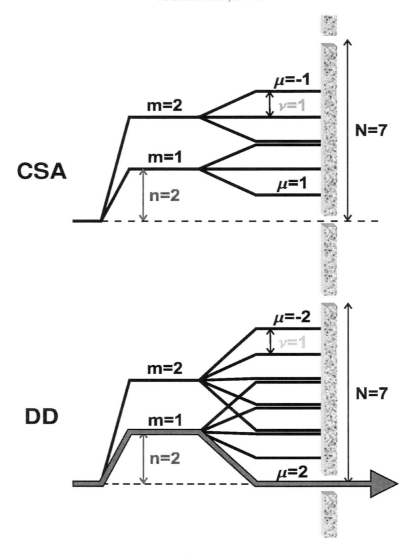

Fig.1 Space-spin selection diagram for $C7_2^1$. The mirror image pathways with m=−1 and m=−2 have been suppressed for the sake of clarity.

<u>*R-type sequences.*</u> The second class of symmetrical rotor-synchronized pulse sequences is based upon 180° rotation elements, denoted R, rather than cycles [7]. One construction procedure for RN_n^v sequences is as follows: (i) take n sample revolution periods; (ii) divide this into N equal intervals; (iii) fill each interval with a rf pulse sequence R, which implements a π rotation around the x-axis, if all other interactions are ignored; (iv) for odd-numbered elements, change the sign of all phase shifts within

the element R; (v) shift the phase of the qth element by the angle $(-1)^q \pi v/N$, where $q=0,1..N-1$.

This construction procedure resembles that for CN_n^v sequences, except for the selection of a 180° rotation element rather than a cycle, the change in the sign of the phase shift for odd-numbered elements, and the alternation of the phases implied by step (v), instead of the incrementation of the phases employed in the C-sequences.

The theory of these sequences will be described elsewhere. The selection rules for RN_n^v sequences are given by [7]:

$$\overline{H}_{lm\lambda\mu}^{(1)} = 0 \text{ if } mn-\mu v \neq (N/2) \times Z_\lambda \tag{2}$$

where Z_λ is an integer of the same parity as the spin rank λ, i.e. if λ is odd, then Z_λ is an odd integer, while if if λ is even, then Z_λ is an even integer, including zero.

The selection rule (2) is more restrictive than the rule (1). In particular, the positions of the holes in the "barrier" depend on the spin rank λ: If λ is even, then the holes occur at levels 0, $\pm N$, $\pm 2N$..; If λ is odd, then the holes occur at levels $\pm N/2$, $\pm 3N/2$.. This dependence on the spin rank λ allows the R-sequences to discriminate between terms with the same values of m and μ, but different values of λ, creating many new possibilities.

The scaling factor. Although the symmetry properties ensure the suppression of undesirable terms in the average Hamiltonian, they say nothing about the magnitude of the terms that are retained. In general, these terms are proportional to a scaling factor, denoted κ, that depends on the detailed structure of the elements C or R. Formulae for the scaling factor are given elsewhere [6]. It is normally desirable that κ is as large as possible for the symmetry-allowed terms.

Supercycles. The performance and robustness of the symmetry-based sequences may be enhanced further by building supercycles. This is done by concatenating the original CN_n^v or RN_n^v sequence with one or more variant cycles, which are usually related to the original one by well-defined "mutations". Some common mutations are: (i) overall phase shifts; (ii) change in sign of the spin winding number v; (iii) cyclic permutation of some pulse sequence elements. The construction of supercycles must be performed with care, as it is easy to destroy some of the favourable properties of the original sequence if the supercycle construction procedure does not respect the appropriate symmetry constraints. Some examples of useful supercycles are given in refs.[3, 5, 6], and in the discussion below.

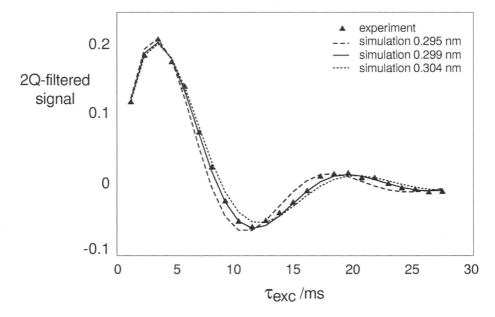

Fig.2. Double-quantum-filtered ^{13}C signal amplitudes in [11,20-$^{13}C_2$]-all-E-retinal, obtained using a R22$_4^9$ sequence at a static field of 9.4 T and a spinning frequency of 7.000 kHz. The triangles are experimental points; the solid line is a numerical simulation for a 0.299 nm distance, multiplied by a biexponential decay function fitted to obtain the best match with experiment (the fitting of the decay has very little impact on the estimated distance). The dashed lines show best fit simulations for distances of 0.295 nm and 0.304 nm. The pulse sequence was as in ref.1, except that a R22$_4^9$ sequence was used instead of C7, and that the 2Q reconversion interval was held fixed at 5.0 ms while the 2Q excitation interval was varied.

Results and Discussion

A brief selection of previously unpublished experimental results for some symmetry-based pulse sequences will now be given. These results are of a preliminary nature and will be published in full elsewhere.

Double-quantum homonuclear recoupling and ^{13}C-^{13}C distance measurements. One of the most important and challenging tasks in magic-angle-spinning NMR is to recouple homonuclear dipolar interactions, allowing the measurement of internuclear distances, for example between ^{13}C nuclei. Double-quantum schemes are particularly important in biomolecular NMR since the special phase properties of excited double-quantum coherences allow the complete suppression of signals from natural abundance ^{13}C spins, even in large molecules. The C7 sequence, which has the symmetry C7$_2^1$ (see Fig.1) has been very successful and is used in a variety of guises [1-3]. However, this symmetry tends to require a rather large ratio of the rf field to the spinning frequency. Recently, we have experimented with different symmetries, such as $C14_4^5$, which may be used in a supercycled form which requires only half the rf field of C7 [6]. Other groups have exploited a supercycled version of $C5_2^1$ symmetry [5]. More recently, we have shown that a large number of promising R-sequence solutions exist [7]. These sequences are often more robust than the CN_n^v sequences, especially with respect to chemical shift anisotropy. Fig.2 shows some experimental results for a $R22_4^9$ sequence, performed on a crystalline sample of $[11,20-^{13}C_2]$-all-E-retinal in which the internuclear distance, as determined by X-ray diffraction, is 0.296 nm [10]. The prominent oscillation is due to the through-space dipole-dipole coupling. The oscillation frequency provides an internuclear distance estimate of 0.299 nm, in close agreement with the X-ray result. The precision of the distance estimate is better than ±3 pm, as may be seen in Fig.2.

We have used $R14_2^6$ and $R22_4^9$ sequences on a number of $^{13}C_2$-labelled compounds. One objective of this work has been to assess the feasibility of accurate C-C bond length measurement in systems with large CSA. Such measurements would be very useful for elucidating the electronic structure of ligands and prosthetic groups, for example the retinylidene chromophore of rhodopsin and its relatives. So far, our results are quite encouraging. We have examined four $^{13}C_2$-labelled compounds, with the reported C-C distances from diffraction measurements spanning the range 0.131 nm to 0.153 nm (the compounds were diammonium $[2,3-^{13}C_2]$-fumarate; ammonium hydrogen $[2,3-^{13}C_2]$-maleate; $[14,15-^{13}C_2]$-all-E-retinal; L-$[2,3-^{13}C_2]$-alanine). This set of compounds includes both double and single C-C bonds and spans a wide range of isotropic shifts and chemical shift anisotropies. In all cases, the $^{13}C-^{13}C$ distances determined by the $R14_2^6$ sequence in a field of 9.4 T agreed with the diffraction methods to within 5 pm. The main source of uncertainty in the NMR distance measurement is the lack of knowledge of the CSA orientations. Despite this, the methods are sufficiently precise to allow informative measurements of C-C bond lengths in biomolecules.

We have also measured longer distances, such as the one shown in Fig.2, as well as the 0.39 nm $^{13}C-^{13}C$ distance in diammonium $[1,4-^{13}C_2]$-fumarate, which was estimated with an accuracy of 10 pm. There is more work to be done, but the method shows promise for investigating the bonding structure at the active site of large, non-crystalline biomolecules such as retinal proteins.

One point of technical importance should be noted: Although the RN_n^v sequences are robust in many respects, they are *exceedingly* sensitive to the accuracy and stability of the rf phase shifts. We often need to adjust the rf phase in steps of 0.1° in order to obtain best performance. Instrument manufacturers take note!

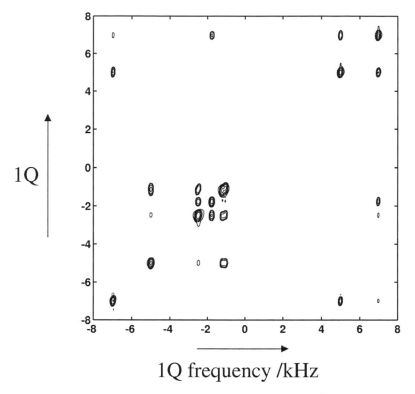

1Q

1Q frequency /kHz

Fig.3. Experimental single-quantum 2D correlation spectrum of L-[U-^{13}C]-tyrosine at a field of 9.4T and a spinning frequency of 23 kHz, obtained using the sequence in Eq.(3), with a duration 1 ms.

<u>*Zero-quantum homonuclear recoupling and ^{13}C-^{13}C correlations.*</u> The symmetry theorems given above may readily be applied to the problem of ZQ homonuclear recoupling. Zero-quantum recoupling sequences may be used to obtain 2D ^{13}C-^{13}C correlations maps, which are useful for spectral assignment. The symmetries may be chosen so as to provide insensitivity to isotropic and anisotropic chemical shifts, unlike the widely-used RFDR sequence, which relies on chemical shift interference in order to work at all [11]. One promising sequence uses the symmetry $R6_6^2$ with the basic element $R = 90_{180}270_0$, incorporated in a 6-fold supercycle. The full sequence has the form

$$[R6_6^2 R6_6^{-2}]_0 [R6_6^2 R6_6^{-2}]_{120} [R6_6^2 R6_6^{-2}]_{240} \tag{3}$$

where $[..]_\phi$ represents an overall phase shift through the angle ϕ. Some experimental results for L-[U-^{13}C]-tyrosine are shown in Fig.3.

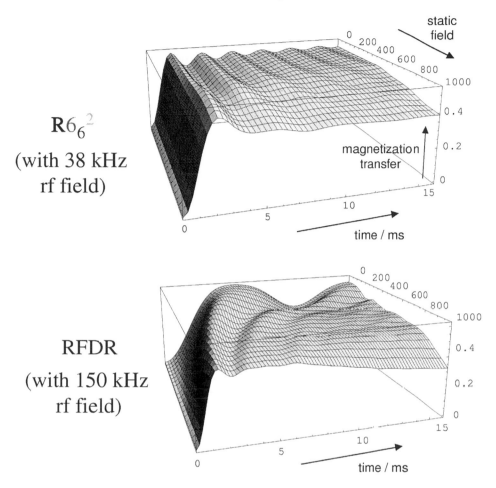

Fig.4. Numerical simulations of powder-average Zeeman magnetization transfer between two ^{13}C spins, at a MAS frequency of 38 kHz. Top: the sequence in Eq.(3); Bottom: RFDR. The magnetization transfer is shown as a function of time (horizontal axis) and static field (specified as the proton Larmor frequency in MHz). The simulation parameters correspond to $[U-^{13}C]$-glycine.

The preliminary numerical simulations shown in Fig.4 indicate that the supercycled $R6_6^2$ sequence should display excellent performance at high magnetic fields (simulated up to 20 T) and high sample spinning frequencies (simulated at 38 kHz), without requiring an excessively large rf field. Its predicted performance is more broadband with respect to chemical shifts and chemical shift anisotropies than RFDR, despite its lower rf field requirements. Furthermore, the new sequence also has much lower rf field requirements than the RIL method [12, 13], which normally cannot be

A. BRINKMANN, ET. AL

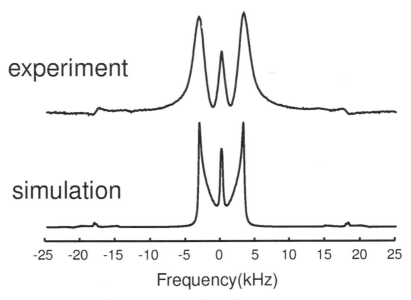

Fig5. (Top) Experimental ^{13}C spectrum of L-[^{15}N, 2-^{13}C]-alanine, obtained by taking the cross-polarized ^{13}C signal in the presence of a R18$_2^5$ sequence at the ^1H Larmor frequency. The basic element of the R18$_2^5$ sequence was R = 180$_0$. The static field was 9.4 T, the spinning frequency was 18 kHz, and the proton nutation frequency was 81 kHz. (Bottom) Numerical simulation using a simple heteronuclear two-spin model, with an effective CH distance of 0.112nm.

implemented at all at high spinning frequencies. In addition, the sequence in Eq.(3) is expected to be tolerant to the setting of the rf phase shift.

Heteronuclear dipolar recoupling with homonuclear decoupling. The symmetry theorems also allow one to design sequences that recouple different spin species, such as ^{13}C and ^1H spins, at the same time as decoupling homonuclear spins from each other [7]. Fig.5 shows the Fourier transform of ^{13}C signals acquired in the presence of R18$_2^5$ irradiation of the protons. The R18$_2^5$ sequence recouples the heteronuclear dipole-dipole interaction while decoupling the protons from each other. The spectral splitting is due to the recoupled heteronuclear dipole-dipole interaction between the ^{13}C spin and its directly-bonded ^1H neighbour. We anticipate that bond lengths and other internuclear distances may simply be read off such one-dimensional spectra, after calibration of the scaling factor κ. This will be useful for investigating hydrogen bonding and other structural issues.

Other symmetries. A variety of symmetries for other decoupling tasks was listed in ref.[7]. We are also preparing a publication on multiple-channel rf sequences (A. Brinkmann, unpublished). An application of C-type sequences to the MAS of oriented systems is sketched by C. Glaubitz elsewhere in this volume.

Existing pulse sequences. The symmetry theorems sketched here also provide insight into the operation of a variety of well-known pulse sequences. For example, the REDOR scheme with xy phase cycling [14-16] conforms to R4$_2^1$ symmetry. This symmetry accounts for the good performance of this sequence, particularly in regimes for which it was not originally designed, such as at fast MAS frequencies [17]. However, the R4$_2^1$ symmetry of REDOR also recouples the irradiated spins with each other -- this undesirable feature may often be unimportant when REDOR is applied to low-γ spins such as ^{15}N, but should be of concern in other situations.

The RFDR homonuclear recoupling method [11], with xy phase cycling, conforms to R4$_4^1$ symmetry. This is an appropriate symmetry for the task of zero-quantum homonuclear recoupling [7]. However, the scaling factor κ of RFDR vanishes in the limit of infinite rf field. The RFDR sequence only works at all because the duration of the rf pulses is finite, and because of assistance from higher-order interference terms involving the chemical shifts. As a result, the magnetization transfer under RFDR is strongly dependent on both the isotropic and anisotropic chemical shifts, as well as the rf field strength. The new symmetry-based sequences, such as the one given in Eq.(3), are expected to be more quantitative and reliable.

The successful TPPM heteronuclear decoupling method [18] is also amenable to symmetry analysis [4, 7].

Conclusions

The symmetry theorems described here allow a wide variety of decoupling and recoupling problems in solid-state NMR to be addressed in a rational and general way. Symmetry does not itself solve all problems, but does provide a sound starting point for designing pulse sequences. We have obtained promising results for double-quantum and zero-quantum homonuclear recoupling sequences, and for heteronuclear recoupling sequences. Some of these solutions are predicted to have good performance at very high magnetic fields and at very fast sample rotation frequencies. Other solutions provide robust and quantitative performance under moderate fields and spinning frequencies, in the presence of large chemical shift anisotropies.

My guess (M.H.L.) is that the future development of solid-state NMR in biology will continue to include several parallel strands, with high and low magnetic fields, high and low spinning frequencies, and a variety of labelling strategies, all playing important roles in different situations. In many cases very high static magnetic fields will lead to good spectral resolution, but these advantages should be balanced against the enhanced effects of the chemical shift anisotropies, which will often compromise the quantitation of methods for geometry determination. Similarly, the use of heavily-labelled compounds increases the information output per sample, but also degrades the quality and reliability of the geometrical information, and increases the complexity of the analysis. For these reasons, solid-state NMR on selectively-labelled compounds at moderate fields will continue to be a sound strategy for obtaining reliable and quantitative geometrical data, at least in many contexts. The biological solid-state NMR spectroscopists of the future will be faced with some interesting strategic choices.

In any case, the design of pulse sequences using symmetry, as sketched here, is versatile enough for a wide range of future scenarios.

Acknowledgements

This research was supported by the Swedish Natural Science Research Council, the Göran Gustafsson Foundation for Research in the Natural Sciences and Medicine, and the European Union Marie Curie Fellowship Program. We would like to thank H. Luthman for synthesis of the labelled retinals, A. Sebald for the loan of other labelled compounds, and Ole Johannessen for experimental help.

References

[1] Lee, Y.K., Kurur, N.D., Helmle, M., Johannessen, O.G., Nielsen, N.C. and Levitt, M.H., Chem. Phys. Lett. 242 (1995) 304.

[2] Hohwy, M., Jakobsen, H.J., Edén, M., Levitt, M.H. and Nielsen, N.C., J. Chem. Phys. 108 (1998) 2686.

[3] Rienstra, C.M., Hatcher, M.E., Mueller, L.J., Sun, B., Fesik, S.W. and Griffin, R.G., J. Am. Chem. Soc. 120 (1998) 10602.

[4] Edén, M. and Levitt, M.H., J. Chem. Phys. 111 (1999) 1511.

[5] Hohwy, M., Rienstra, C.M., Jaroniec, C.P. and Griffin, R.G., J. Chem. Phys. 110 (1999) 7983.

[6] Brinkmann, A., Edén, M. and Levitt, M.H., J. Chem. Phys. 112 (2000) 8539.

[7] Carravetta, M., Edén, M., Zhao, X., Brinkmann, A. and Levitt, M.H., Chem. Phys. Lett. 321 (2000) 205.

[8] Tycko, R. and Dabbagh, G., J. Am. Chem. Soc. 113 (1991) 9444.

[9] Nielsen, N.C., Bildsøe, H., Jakobsen, H.J. and Levitt, M.H., J. Chem. Phys. 101 (1994) 1805.

[10] Hamanaka, T., Mitsui, T., Ashida, T. and Kakudo, M., Acta Cryst. B28 (1972) 214.

[11] Bennett, A.E., Ok, J.H., Griffin, R.G. and Vega, S., J. Chem. Phys. 96 (1992) 8624.

[12] Baldus, M., Tomaselli, M., Meier, B.H. and Ernst, R.R., Chem. Phys. Lett. 230 (1994) 329.

[13] Baldus, M., Geurts, D.G. and Meier, B.H., Sol. State Nucl. Magn. Reson. 11 (1998) 157.

[14] Gullion, T. and Schaeffer, J., J. Magn. Reson. 81 (1989) 196.

[15] Garbow, J.R. and Gullion, T., J. Magn. Reson. 95 (1991) 442.

[16] Holl, S.M., Marshall, G.R., Beusen, D.D., Kociolek, K., Redlinski, A.S., Leplawy, M.T., McKay, R.A., Vega, S. and Schaefer, J., J. Am. Chem. Soc. 114 (1992) 4830.

[17] Jaroniec, C.P., Tounge, B.A., Rienstra, C.M., Herzfeld, J. and Griffin, R.G., J. Magn. Reson. 146 (2000) 132.

[18] Bennett, A.E., Rienstra, C.M., Auger, M., Lakshmi, K.V. and Griffin, R.G., J. Chem. Phys. 103 (1995) 1.

Accurate [13]C-[15]N Distance Measurements in Uniformly ^{13}C,[15]N-Labeled Peptides

C.P. Jaroniec[a], B.A. Tounge[b], J. Herzfeld[b], and R.G. Griffin[a]

[a]*Department of Chemistry and Francis Bitter Magnet Laboratory,
Massachusetts Institute of Technology, Cambridge, MA 02139, USA*
[b]*Department of Chemistry, Brandeis University, Waltham, MA 02254, USA*

Introduction

The ability to accurately measure [13]C-^{15}N dipolar couplings corresponding to internuclear distances in the 3-6 Å regime is important for constraining the three-dimensional structure of biological solids. Solid-state NMR (SSNMR) methods for heteronuclear distance measurements in isolated spin pairs are now well-established and will continue to provide valuable structural information [1,2]. However, these methods require synthesis of molecules isotopically labeled in a pairwise fashion, which can be both laborious and expensive. Thus, there is a clear motivation for the development of analogous SSNMR methods for larger spin systems, where multiple internuclear distances can be determined [3-5]. However, in multispin systems [13]C-^{15}N distance measurements have the potential to be complicated by the presence of multiple homonuclear and heteronuclear spin-spin couplings.

In the following we describe a magic angle spinning (MAS) NMR experiment for selective recoupling of [13]C-^{15}N dipolar interactions in uniformly ^{13}C,[15]N-labeled solids [6] and use it to measure multiple [13]C-^{15}N distances in the 3-6 Å range in a U-^{13}C,[15]N-labeled tripeptide, N-formyl-L-Met-L-Leu-L-Phe.

Methods

The selective recoupling of [13]C-^{15}N interactions in a U-^{13}C,[15]N-labeled system is accomplished by the pulse sequence shown in Fig. 1. The experiment employs simultaneous frequency selective π pulses applied to the [13]C-^{15}N spin pair of interest,

S.R.Kiihne and H.J.M.deGroot (ed.), Perspectives on Solid State NMR in Biology, 15-21.

C.P. JARONIEC, ET AL.

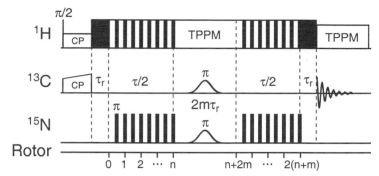

Fig. 1. Frequency selective REDOR pulse sequence. Ramped CP [7] creates the initial ^{13}C magnetization and ^{13}C-^{15}N couplings are reintroduced in a nonselective fashion using REDOR [8]. Selective recoupling of ^{13}C-^{15}N spin pairs is achieved by the generation of a ^{13}C spin echo in a frequency selective fashion using a pair of rotor-synchronized Gaussian π pulses. Couplings to ^1H during the ^{13}C-^{15}N recoupling period are attenuated using a combination of CW and TPPM [9] decoupling (represented by filled and hollow rectangles, respectively). Reference (S_0) experiments are obtained by acquiring spectra in the absence of the ^{15}N selective pulse.

bracketed by two identical periods of nonselective dipolar recoupling. We assume: (i) heteronuclear recoupling sequences with an effective dipolar Hamiltonian of the form $H_D = \sum_{i,j} \omega_D^{ij} 2 C_{iz} N_{jz}$ (where the C and N operators refer to ^{13}C and ^{15}N spins, respectively, and ω_D^{ij} is the orientation dependent C_i-N_j dipolar coupling), and (ii) the weak coupling limit for ^{13}C-^{13}C J couplings ($H_J = 2\pi J_{CC} C_{1z} C_{2z}$).

It can be shown [6] that the observable ^{13}C signal for the complete pulse sequence with the selective pulses applied to spins C_k and N_l is given by [8]:

$$S(\tau) \propto \int_0^\pi d\beta \sin(\beta) \int_0^{2\pi} d\gamma \cos\left(\omega_D^{kl}\tau\right), \qquad (1)$$

where τ is the total dipolar evolution time (see Fig. 1), ω_D^{kl} is the effective C_k-N_l dipolar coupling (*vide infra*), which depends on the Euler angles, β and γ, defining the orientation of the dipolar tensor in the rotor-fixed reference frame (the integrals indicate the average over all orientations of the dipolar tensor in the powder sample). In other words, the ^{13}C-^{15}N dipolar coupling for the spin pair irradiated by both selective pulses is retained, while all remaining ^{13}C-^{15}N dipolar couplings and ^{13}C-^{13}C J couplings to the ^{13}C spin of interest are refocused [6,10]. In the present implementation of the experiment REDOR [8] was employed for nonselective heteronuclear recoupling since it is well-compensated for pulse imperfections. Hence, we refer to the new experiment as *frequency selective REDOR* (FSR).

Since high-field studies of U-^{13}C,^{15}N-labeled systems normally employ high MAS frequencies ($\omega_r/2\pi \approx 10$-30 kHz), we briefly consider the performance of REDOR

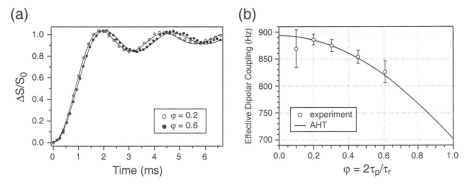

Fig. 2. Finite pulse effects during REDOR. (a) Experimental REDOR $\Delta S/S_0$ curves for [2-$^{13}C,^{15}N$]glycine recorded with: (◦) φ=0.2 ($\omega_r/2\pi$=10 kHz, $\omega_{rf}(^{15}N)/2\pi$=50 kHz) and (•) φ=0.6 ($\omega_r/2\pi$=15.152 kHz, $\omega_{rf}(^{15}N)/2\pi$=25 kHz). Simulations are shown as solid lines with best-fit effective dipolar couplings of 886±10 Hz for φ=0.2 and 826±20 Hz for φ=0.6. (b) Comparison of the effective dipolar couplings determined for [2-$^{13}C,^{15}N$]glycine as a function of φ (◦) and the theoretical curve (solid line) obtained using average Hamiltonian theory (cf. Eq. 3). All experiments were performed at 11.7 T. The spinning frequency was varied between 5.0 and 15.152 kHz (controlled to ±5 Hz) and ^{15}N rf fields were 25-50 kHz. The ^{1}H decoupling was 83 kHz CW during REDOR and 83 kHz TPPM [9] during acquisition. The ^{13}C refocusing pulse was 10 μs. [2-$^{13}C,^{15}N$]glycine (not diluted in natural abundance material) was recrystalized from aqueous solution, and experiments were performed at room temperature.

under conditions where the rf irradiation occupies a significant fraction of the rotor period. For the REDOR sequence shown in Fig. 1, where all π pulses are applied to the non-observed (^{15}N) spins and phase-alternated according to xy-4 or extensions thereof [11], the first-order effective Hamiltonian for the C-N dipolar coupling is given by [12]

$$H_D^{(1)} = \omega_D 2C_z N_z \tag{2}$$

with

$$\omega_D = -\frac{\sqrt{2}}{\pi} b_{CN} \frac{\cos\left(\frac{\pi}{2}\varphi\right)}{1-\varphi^2} \sin(2\beta)\sin(\gamma). \tag{3}$$

The C-N dipolar coupling constant, b_{CN} is proportional to the inverse cube of the internuclear distance, r_{CN}, and φ is the fraction of the rotor period, τ_r, occupied by ^{15}N π pulses of length τ_p:

$$\varphi = \frac{2\tau_p}{\tau_r}. \tag{4}$$

We note that Eq. 3 differs from the expression for ideal δ-pulse REDOR [8] only by the scaling factor $\cos(\varphi\pi/2)/(1-\varphi^2)$. This means that b_{CN} can be scaled down by as much as π/4 due to finite ^{15}N π pulses. However, for most experimental conditions of practical interest the scaling factor remains close to unity.

Results and Discussion

In Fig. 2 we investigate the finite pulse effects on REDOR experiments in a model two-spin system, $[2\text{-}^{13}C,^{15}N]$glycine, where the fraction of the rotor period occupied by the rf pulses was varied between ~ 0.1 and 0.6. As predicted by average Hamiltonian theory (AHT) [13], a minor scaling (<10%) of the dipolar oscillation frequency is observed as φ increases from 0.1 to 0.6 and the overall agreement between the experiments and AHT is good.

Having considered the performance of REDOR at high MAS frequencies required for our experiments, we proceed with the discussion of the FSR experiment. In Fig. 3 we qualitatively investigate the selective recoupling of $C^\beta\text{-}N'$ and $C^\beta\text{-}N^{\delta2}$ interactions in $[U\text{-}^{13}C,^{15}N]$asparagine. (Asparagine represents a favorable model system for selective distance measurements in multispin clusters since the C^α and C^β as well as N' and $N^{\delta2}$ resonances are well-resolved (spectra not shown).) A 1.0 ms Gaussian pulse is applied to the C^β resonance and ^{15}N Gaussian pulses are applied at different

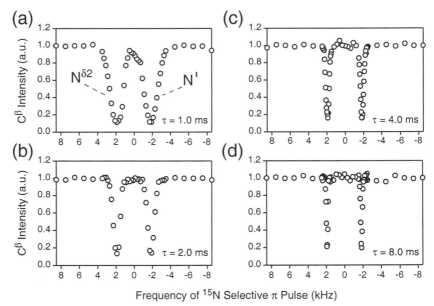

Fig. 3. Selective recoupling of $C^\beta\text{-}N'$ and $C^\beta\text{-}N^{\delta2}$ dipolar interactions in $[U\text{-}^{13}C,^{15}N]$asparagine. The FSR pulse sequence shown in Fig. 1 was used with the total dipolar evolution time $\tau = 8.0$ ms. All experiments were performed at $\omega_r/2\pi=10$ kHz $\pm$ 5 Hz with a 1.0 ms Gaussian pulse applied to the C^β resonance. In (a)-(d) the results of 'frequency sweep' experiments are shown for ^{15}N Gaussian pulses of: (a) 1.0, (b) 2.0, (c) 4.0 and (d) 8.0 ms. The bandwidths outside of which negligible recoupling occurs were found to be approximately: (a) ±2000, (b) 1000, (c) 500 and (d) 250 Hz. Each experimental point (∘) corresponds to the ratio of the intensity of the C^β resonance in the presence and absence of the ^{15}N Gaussian pulse.

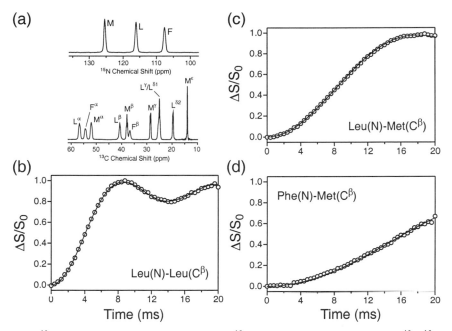

Fig. 4. (a) ^{13}C (only aliphatic region shown) and ^{15}N MAS spectra of N-formyl-[U-^{13}C,^{15}N]Met-Leu-Phe recorded at $\omega_r/2\pi=10$ kHz. In (b)-(d) representative internuclear distance measurements in N-formyl-[U-^{13}C,^{15}N]Met-Leu-Phe are shown. Experimental $\Delta S/S_0$ curves (○) and analytical simulations (solid lines) are shown for selective recoupling of: (a) Leu(N)-Leu(C$^\beta$), (b) Leu(N)-Met(C$^\beta$) and (c) Phe(N)-Met(C$^\beta$). Experiments were performed at $\omega_r/2\pi=10$ kHz ± 5 Hz and 32 scans were averaged for each time point. ^{13}C and ^{15}N Gaussian pulses were 2.0 and 10.0 ms, respectively. The experiments shown here were performed on N-formyl-[U-^{13}C,^{15}N]Met-Leu-Phe undiluted in the natural abundance peptide. The summary of all distance measurements in a diluted N-formyl-[U-^{13}C,^{15}N]Met-Leu-Phe sample is given in Fig. 5.

frequencies in the ^{15}N spectrum. Fig. 3 demonstrates the selective recoupling of C$^\beta$-N' and C$^\beta$-N$^{\delta 2}$ interactions using the frequency selective spin-echo generated by the two selective pulses. The C$^\beta$ signal is dephased only by the ^{15}N spins on-resonance with the Gaussian pulse, while couplings to off-resonance nuclei are refocused by the pulse sequence. The spectral selectivity of the FSR experiment can be tuned by adjusting the duration of the ^{15}N Gaussian pulse (e.g., the recoupling bandwidth obtained with a 8 ms pulse is ~ ±250 Hz).

Representative FSR distance measurements in N-formyl-[U-^{13}C,^{15}N]Met-Leu-Phe (MLF) (sample not diluted in natural abundance peptide) are shown in Fig. 4. The ^{13}C and ^{15}N MAS spectra of MLF (Fig. 4a) suggest that distance measurements between most sidechain ^{13}C and backbone ^{15}N should be feasible. Indeed, the use of ^{15}N pulses in the 6-10 ms range enables the selective ^{13}C-^{15}N recoupling, and in a [U-^{13}C,^{15}N]MLF sample diluted to ~10% in natural abundance MLF we have measured 15 sidechain to

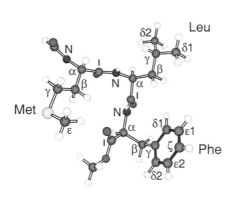

atoms		NMR (Å)	X-ray (Å)
M(N)	M(C$^\beta$)	2.52 +/- 0.02	2.50
	M(C$^\gamma$)	3.20 +/- 0.03	3.04
	M(C$^\epsilon$)	5.4 +/- 0.3	5.71
	L(C$^\beta$)	5.7 +/- 0.7	6.03
	L(C$^{\delta 2}$)	5.5 +/- 0.3	6.28
L(N)	M(C$^\beta$)	3.12 +/- 0.03	3.20
	M(C$^\gamma$)	4.17 +/- 0.10	4.56
	M(C$^\epsilon$)	5.5 +/- 0.3	5.93
	L(C$^\beta$)	2.46 +/- 0.02	2.50
	L(C$^{\delta 2}$)	3.64 +/- 0.09	3.63
F(N)	M(C$^\beta$)	4.12 +/- 0.15	4.06
	M(C$^\gamma$)	4.8 +/- 0.2	5.43
	M(C$^\epsilon$)	5.2 +/- 0.3	5.62
	L(C$^\beta$)	3.24 +/- 0.12	3.12
	L(C$^{\delta 2}$)	5.4 +/- 0.3	5.38

Fig. 5. X-ray structure of N-formyl-Met-Leu-Phe-OMe [14] and the summary of internuclear distances in N-formyl-[U-^{13}C,^{15}N]Met-Leu-Phe determined using frequency selective REDOR. The distances were measured in a N-formyl-[U-^{13}C,^{15}N]Met-Leu-Phe sample diluted to ~10% in natural abundance peptide to minimize intermolecular ^{13}C-^{15}N couplings (uncertainties are reported at the 95% level). For comparison with the NMR results, we include the corresponding X-ray distances in N-formyl-Met-Leu-Phe-OMe.

backbone ^{13}C-^{15}N distances in the 3-6 Å range with the precision of ~ 0.1-0.3 Å [6]. The distance measurements are summarized in Fig. 5. Since no X-ray structure has been reported for N-formyl-L-Met-L-Leu-L-Phe, the structure of its methyl ester analogue, N-formyl-L-Met-L-Leu-L-Phe-OMe [14] is shown for comparison. We note here that a set of solid state NMR constraints for N-formyl-Met-Leu-Phe has been compiled by combining the ^{13}C-^{15}N distances determined above with multiple ϕ, ψ and χ dihedral angle constraints [15]. The calculation of the three-dimensional structure of the peptide based on these constraints is currently in progress [16].

Conclusions

To date most solid state NMR structural studies have employed methods applicable to molecules containing labeled spin pairs (^{13}C-^{13}C, ^{13}C-^{15}N, etc.). Furthermore, it has been assumed that certain of these methods function best at slow spinning frequencies where resolution is compromised. Here we have shown that it is possible to perform heteronuclear recoupling experiments such as REDOR in the rapid MAS regime, and to accurately measure multiple ^{13}C-^{15}N distances in uniformly ^{13}C,^{15}N-labeled molecules. In the studies of the tripeptide N-formyl-Met-Leu-Phe we have performed accurate measurements of 15 distances that constrain the structure of the molecule. This is one of the initial investigations to achieve this goal, and we anticipate that the approaches described here will be applicable to larger systems, and will stimulate further developments along these lines.

Acknowledgements

This research was supported by the National Institutes of Health (GM-23403, GM-23289, GM-36810, AG-14366, and RR-00995. C.P.J. is the National Science Foundation Predoctoral Fellow and B.A.T. is the American Cancer Society Postdoctoral Fellow. We thank C.M. Rienstra, M. Hohwy, and B. Reif for stimulating discussions and D.J. Ruben, A. Thakkar and P. Allen for technical assistance.

References

[1]   Bennett, A. E.; Griffin, R. G.; Vega, S. *Recoupling of homo- and heteronuclear dipolar interactions in rotating solids*; Blumich, B., Ed.; Springer-Verlag: Berlin, 1994; Vol. 33, pp 1-77.

[2]   Dusold, S.; Sebald, A. *Annu. Rep. Nucl. Magn. Reson. Spectr. 41* (2000) 185-264.

[3]   Schaefer, J. *J. Magn. Reson. 137* (1999) 272-275.

[4]   Gullion, T.; Pennington, C. H. *Chem. Phys. Lett. 290* (1998) 88-93.

[5]   Liivak, O.; Zax, D. B. *J. Chem. Phys. 113* (2000) 1088-1096.

[6]   Jaroniec, C. P.; Tounge, B. A.; Herzfeld, J.; Griffin, R. G. *J. Am. Chem. Soc.* (2001) in press.

[7]   Metz, G.; Wu, X.; Smith, S. O. *J. Magn. Reson. A 110* (1994) 219-227.

[8]   Gullion, T.; Schaefer, J. *Adv. Magn. Reson. 13* (1989) 57-83.

[9]   Bennett, A. E.; Rienstra, C. M.; Auger, M.; Lakshmi, K. V.; Griffin, R. G. *J. Chem. Phys. 103* (1995) 6951-6957.

[10]   Jaroniec, C. P.; Tounge, B. A.; Rienstra, C. M.; Herzfeld, J.; Griffin, R. G. *J. Am. Chem. Soc. 121* (1999) 10237-10238.

[11]   Gullion, T.; Baker, D. B.; Conradi, M. S. *J. Magn. Reson. 89* (1990) 479-484.

[12]   Jaroniec, C. P.; Tounge, B. A.; Rienstra, C. M.; Herzfeld, J.; Griffin, R. G. *J. Magn. Reson. 146* (2000) 132-139.

[13]   Haeberlen, U. *High-Resolution NMR in Solids: Selective Averaging*; Academic Press: New York, 1976.

[14]   Gavuzzo, E.; Mazza, F.; Pochetti, G.; Scatturin, A. *Int. J. Peptide Protein Res. 34* (1989) 409-415.

Selectivity of Double-Quantum Filtered Rotational-Resonance Experiments on Larger-than-Two-Spin Systems

Matthias Bechmann, Xavier Helluy, and Angelika Sebald
Bayerisches Geoinstitut, Universität Bayreuth, 95440 Bayreuth, Germany

Introduction

Characterizing the orientation and molecular conformation of small organic molecules bound to the inner or outer surfaces of proteins represents an important step in drug design and in understanding the mechanisms of biochemical reactions, and similarly, of non-biological catalytic reactions. In a biochemical context, such molecular units or subunits may often contain only three or four carbon atoms, examples being the pyruvate anion, fumaric and maleic acid derivatives, or the phosphenolpyruvate moiety in differing degrees of ionization. Magic-angle spinning (MAS) NMR experiments, capable of delivering reliable information about the conformational properties of these molecular units, have to combine several properties in order to be able to fulfill these tasks in realistic application situations. First, the ^{13}C resonances originating from the (fully or partially) ^{13}C enriched substrate molecules of interest have to be separable from additional natural-abundance ^{13}C resonances; this calls for the application of double-quantum filtration (DQF) techniques. Second, many of these small substrate molecules feature structural subunits that require using the orientation dependence of ^{13}C chemical shielding as the source of information about molecular conformation; this calls for MAS NMR experiments where magnitudes and orientations of chemical shielding tensors are sensitively reflected. Third, for reasons of synthetic feasibility, the chosen MAS NMR techniques must be applicable in a quantifiable manner to larger-than-two-spin systems. The ease and robustness of the experimental and numerical implementations are an additional consideration.

With these selection criteria in mind, we turn to the so-called rotational-resonance (R^2) condition [1-5] in conjunction with double-quantum filtration (DQF). In the context of larger-than-two-spin systems, a certain preserved narrowbandedness of (some of) the R^2 condition(s) can be at an advantage over more broadbanded alternatives, such as the DQ-DRAWS experiment [6] or the C7 sequence [7] and its

S.R.Kühne and H.J.M.deGroot (eds.), Perspectives on Solid State NMR in Biology, 23-31.

derivatives [8]. We will employ a recently introduced R^2-DQF pulse sequence [9] to investigate aspects of selectivity when applying R^2-DQF experiments to spin systems composed of more than two ^{13}C spins. We use the ^{13}C-three spin system in fully ^{13}C enriched sodium pyruvate, 1-U^{13}C, as our model case. 1-U^{13}C was chosen because i) the

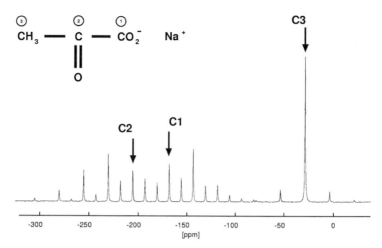

Fig. 1: ^{13}C MAS NMR spectrum of sodium pyruvate ($\omega_0/2\pi = -75.5$ MHz; $\omega_r/2\pi = 1888$ Hz) with ^{13}C in natural abundance; the assignment of the three ^{13}C isotropic chemical shielding values is indicated.

crystal structure of sodium pyruvate is known [10], ii) the parameters of its ^{13}C three-spin system have been determined [11], and iii) this spin system makes a range of rather different R^2 conditions accessible, as can be seen in Fig. 1, where a ^{13}C MAS NMR spectrum of sodium pyruvate with ^{13}C in natural abundance is depicted.

Methods

Fully ^{13}C-enriched sodium pyruvate, 1-U^{13}C, is commercially available (ISOTEC Inc., USA) and was used as received. ^{13}C R^2-DQF experiments at ^{13}C Larmor frequency $\omega_0/2\pi = -50.3$ MHz were run on a Bruker MSL 200 NMR spectrometer using a 4 mm double-bearing CP MAS probe. A range of ^{13}C R^2-DQF experiments on 1-U^{13}C and 1-U$^{13}C_{dil}$ (sample diluted by co-crystallization with ^{13}C natural abundance material in a 1:5 enriched:unenriched ratio) yielded identical spectral lineshapes.

The pulse sequence used for DQF at the $n = 1$ R^2 condition [9] is depicted in Fig. 2. Experimentally ^{13}C $\pi/2$ pulse durations of 3.5 μs and a c.w. ^1H decoupling amplitude of 83 kHz were employed.

The parameters of the ^{13}C spin system in solid sodium pyruvate [11] and a full description of the notation, definitions, and numerical simulation methods used are given elsewhere. [12]

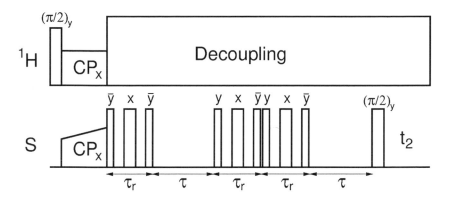

Fig. 2: Pulse sequence to achieve DQF at the n = 1 R^2 condition, where τ_r denotes rotation period, and the three-pulse subsequences consist of $\pi/4$-$\pi/2$-$\pi/4$ pulses [9].

Results and Discussion

For the R^2-DQF pulse sequence ([9], see Fig. 2) it has been demonstrated that high R^2-DQF efficiencies are achieved for large and small dipolar coupling interactions, provided that the chemical shielding anisotropies (csa; δ^{CS}) are substantially less than the difference in the isotropic chemical shielding, $n\omega_r = \omega_{iso}^\Delta \times \delta^{CS_{i,j}}$, with n being a small integer [11]. In the presence of large CSA's, another R^2 DQF pulse sequence [13] maintains higher R^2-DQF efficiencies.

The ^{13}C three-spin system in **1-U^{13}C** presents a set of three, rather different $n = 1$ R^2 conditions. When choosing the ^{13}C2-^{13}C3 pair, $\omega_{iso}^{\Delta_{23}} = \omega_r$ substantially exceeds the magnitude of all spin interactions in **1-U^{13}C**. The ^{13}C1-^{13}C3 pair is similarly characterized by a fairly large value $\omega_{iso}^{\Delta_{13}}$, but features a much smaller dipolar coupling constant b_{13} than the ^{13}C2-^{13}C3 pair. The $n = 1$ R^2 condition for the ^{13}C1-^{13}C2 pair in **1-U^{13}C** differs strongly: these two ^{13}C spins are characterized by a small value of $\omega_{iso}^{\Delta_{12}}$, by substantial chemical shielding anisotropies, $\delta^{CS_{12}}$, and by a large value of b_{12}. At a ^{13}C Larmor frequency of $\omega_0 / 2\pi = -50.3$ MHz, $\delta^{CS_{12}}$ (considerably) and b_{12} (slightly) exceed ω_r. A comparison of theoretically expected (—) and experimentally observed (o) R^2-DQF efficiencies, plotted as a function of τ, for these three $n = 1$ R^2 conditions in **1-U^{13}C** is shown in Fig. 3.

The trends in R^2-DQF efficiencies for **1-U^{13}C** follow the expectations based on previous investigations of pairwise selectively ^{13}C2,^{13}C3 [11] and ^{13}C1,^{13}C2 [12] isotopomers of sodium pyruvate. The pulse sequence depicted in Fig. 2 yields fairly

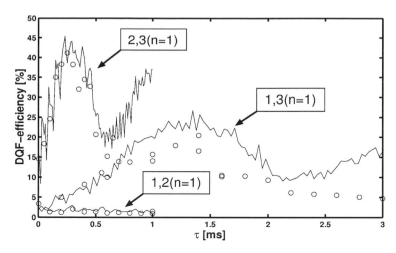

Fig. 3: Theoretically expected (—) and experimentally observed (o) n = 1 R^2-DQF efficiencies in 1-U^{13}C, plotted as a function of τ. The simulations employ the known parameters of this ^{13}C three-spin system [11]; the individually chosen i,j R^2 conditions are indicated. The experimental data were obtained at $\omega_0 / 2\pi$ = –50.3 MHz, with $\omega_r / 2\pi$ = 1832 Hz (^{13}C1,^{13}C2 pair), $\omega_r / 2\pi$ = 7020 Hz (^{13}C1,^{13}C3 pair), and $\omega_r / 2\pi$ = 8882 Hz (^{13}C2,^{13}C3 pair). The efficiency is given in percent with the integrated spectral intensity of the chosen i,j pair in the corresponding conventional R^2 spectrum taken as 100 percent.

high to high efficiences for small and large dipolar coupling constants at R^2 conditions where ω_r considerably exceeds the chemical shielding anisotropies present. In the presence of substantial chemical shielding anisotropies, other sequences [13] yield higher efficiencies and offer a more suitable experimental route to the determination of chemical shielding tensor orientations from R^2-DQF lineshapes [12].

The R^2-DQF efficiency curves for the ^{13}C2,^{13}C3 pair in 1-U^{13}C follow very closely the corresponding curves for the pairwise selectively ^{13}C2,^{13}C3 enriched isotopomer; in addition, the experimentally observed R^2-DQF lineshapes for this spin pair in the two isotopomers were found to be indistinguishable [11]. This is further corroborated by numerical simulations employing three-spin calculations (see Fig. 4 a) or two-spin simulations (see Figure 4 b). The large difference in isotropic chemical shielding $\omega_{iso}^{\Delta 23}$ in conjunction with the ^{13}C2,^{13}C3 n = 1 R^2(-DQF) condition reduces the three-spin system in 1-U^{13}C to an effective ^{13}C2,^{13}C3 two-spin system. This simplification is accompanied by a reduced information content: under these specific n = 1 R^2 and R^2-DQF conditions, only the magnitude of the dipolar coupling constant b_{23} is sensitively reflected in the resulting lineshapes.

Analogously, the situation for the ^{13}C1,^{13}C3 n = 1 R^2-DQF lineshapes of 1-U^{13}C is now examined more closely, addressing the situation where $\omega_{iso}^{\Delta 13} < \omega_{iso}^{\Delta 23}$, and $b_{13} \times$

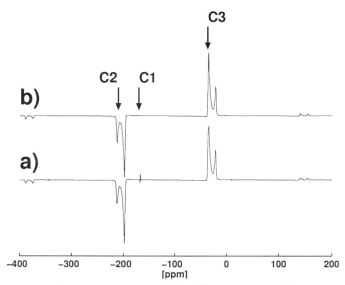

Fig. 4: Simulated n = 1 R^2-DQF spectra for the $^{13}C2,^{13}C3$ pair in 1-$U^{13}C$, with $\omega_0 / 2\pi = -50.3$ MHz, $\omega_r / 2\pi = 8882$ Hz, $\tau = 250$ μs, employing the known parameters of the pyruvate ^{13}C spin system. a): full three-spin simulation; b): two-spin simulation ignoring $^{13}C1$.

b_{23}. The smaller value $b_{13} / 2\pi = -430$ Hz does not dramatically reduce the R^2-DQF efficiency since $\omega_{iso}^{\Delta_{13}} > \delta^{CS_1}$; but the slightly reduced overall efficiency as compared to the previous $^{13}C2,^{13}C3$ case does arise as a function of the now slightly increased 'relative weight' of δ^{CS_1} in relation to $\omega_{iso}^{\Delta_{13}}$ as compared to the δ^{CS_2} to $\omega_{iso}^{\Delta_{23}}$ ratio. Of course, the smaller value $b_{13} / 2\pi = -430$ Hz is reflected in less pronounced splittings of the $^{13}C1,^{13}C3$ selected $n = 1$ R^2-DQF lineshapes of 1-$U^{13}C$. An experimental $^{13}C1,^{13}C3$ selected $n = 1$ R^2-DQF spectrum is shown in Fig. 5 a, in comparison with the corresponding simulated spectrum employing a three-spin simulation in Fig. 5 b. The two lineshapes agree quite well. Describing the spectrum by a $^{13}C1,^{13}C3$ two-spin simulation with the known parameters of the two spins does not give acceptable agreement between experimentally measured and simulated lineshapes. Extending the $^{13}C1,^{13}C3$ two-spin simulation to iterative fitting with b_{13} as a free fit parameter eventually leads to good agreement between experimental and best-fit simulated lineshapes (see Fig. 5 c). However, then the two-spin best-fit value found for b_{13} is -510 Hz. In other words: treating the $^{13}C1,^{13}C3$ $n = 1$ R^2-DQF lineshapes of 1-$U^{13}C$ as originating from a $^{13}C1,^{13}C3$ two-spin system, underestimates the $^{13}C1$-$^{13}C3$ internuclear distance as being 246 pm, compared to the known value of 260.5 pm. Similar deviations are found when using other experimental $^{13}C1,^{13}C3$ selected $n = 1$ R^2-DQF lineshapes of 1-$U^{13}C$ as input for simulations and iterative lineshape fits. There

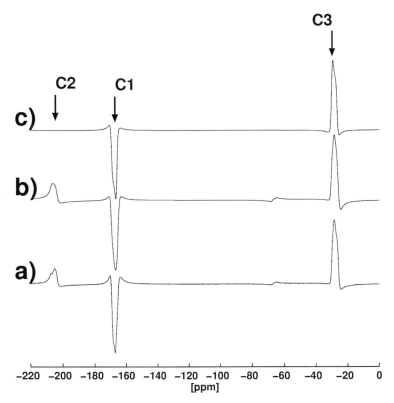

*Fig. 5: n = 1 R^2-DQF spectra for the $^{13}C1,^{13}C3$ pair in **1-U^{13}C**, with $\omega_0 / 2\pi$ = −50.3 MHz, ω_r / 2π = 7020 Hz, τ = 400 μs, and employing the known parameters of the pyruvate ^{13}C spin system in the simulations. a): experimental spectrum; b): three-spin simulation; c): best-fit simulation with a $^{13}C1,^{13}C3$ two-spin approximation, corresponding to $b_{13} / 2\pi$ = −510 Hz. The simulations shown in b) and c) employ the known Euler angles $\Omega_{PC'}^{CS_{1,3}}$.*

may well be applications where this approximation would appear as sufficiently accurate.

A completely different situation is encountered with the R^2-DQF spectra of **1-U^{13}C** adjusted for the $^{13}C1,^{13}C2$ $n = 1$ R^2 condition, with δ^{CS_1} = 2.20 $\omega_{iso}^{\Delta_{12}}$ and δ^{CS_2} = 2.95 $\omega_{iso}^{\Delta_{12}}$. At $\omega_0 / 2\pi$ = −50.3 MHz, the appropriate MAS frequency $\omega_r / 2\pi$ = 1832 Hz is slightly less than the dipolar coupling constants b_{12} and b_{23}, and is fairly close to the $n = 4$ $^{13}C1,^{13}C2$ and $n = 5$ $^{13}C2,^{13}C3$ R^2 conditions, respectively. Figure 6 illustrates the properties of the $^{13}C1,^{13}C2$ $n = 1$ selected R^2-DQF spectra of **1-U^{13}C**.

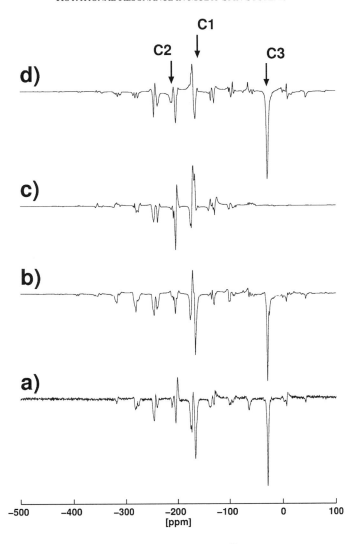

*Fig. 6: n = 1 R^2-DQF spectra for the $^{13}C1$,$^{13}C2$ pair in **1-U^{13}C**, with $\omega_0 / 2\pi = -50.3$ MHz, $\omega_r / 2\pi = 1832$ Hz, $\tau = 700$ μs. a): experimental spectrum; b): three-spin simulation based on the known parameters from R^2 spectra [11]; c): same, but $^{13}C1$,$^{13}C2$ two-spin simulation; d): three-spin simulation based on the known parameters of the spin system, but orientation of the ^{13}C chemical shielding tensors changed from the correct values $\Omega_{PC}^{CS_1} = \{135,0,0\}$, $\Omega_{PC}^{CS_2} = \{0,95,90\}$ to $\Omega_{PC\ assumed}^{CS_1} = \{180,90,0\}$, $\Omega_{PC\ assumed}^{CS_2} = \{0,45,0\}$.*

An experimental spectrum, obtained with $\tau = 0.7$ ms is shown in Fig. 6 a, the corresponding simulated spectrum is displayed in Fig. 6 b. Agreement of the two lineshapes is fairly good, though with some room for improvement: the spin system parameters of 1-$U^{13}C$ had previously been determined by iterative lineshape fitting of conventional R^2 spectra; it has been shown that csa orientational parameters are more sensitively reflected in R^2-DQF lineshapes than in the corresponding R^2 spectra [12]. Clearly, describing this R^2-DQF spectrum of 1-$U^{13}C$ by a $^{13}C1,^{13}C2$ two-spin approximation is an invalid approximation (see Fig. 6 c). The simulated R^2-DQF spectrum in Fig. 6 d illustrates that changes in the Euler angles $\Omega_{PC}^{CS_{1,2}}$, describing the orientations of the chemical shielding tensor orientations, are sensitively reflected in the R^2-DQF lineshapes. Depending on the kind of information one is aiming to extract, one may consider the 'all included' character of these $^{13}C1,^{13}C2$ $n = 1$ selected R^2-DQF spectra of 1-$U^{13}C$ as a blessing or a curse. It is a blessing if, for instance, one wants to determine the absolute orientations of the chemical shielding tensors in a three-spin system from as few experimental spectra as possible. It is a curse if the main interest is focussed on the $^{13}C1,^{13}C2$ pair itself. Then, however, it would be straightforward to emphasize the $^{13}C1,^{13}C2$ two-spin character of these spectra, simply by running similar experiments at a (much) higher Larmor frequency.

Conclusions

A protocol that combines R^2-DQF experiments [9,13] with iterative lineshape fitting approaches, based on numerically exact simulations should be capable of delivering complete information on the geometry of small, isolated molecules or molecular fragments in nearly unrestricted circumstances. With only minimal advance knowledge of the spin-system properties, it is possible to predefine a suitable set of three to four different R^2-DQF experiments (pulse sequence, R^2 order, and/or Larmor frequency). Since the degree of selectivity of the various R^2-DQF experiments can be tailored to some extent by the choice of the experimental R^2 conditions, a small set of one-dimensional R^2-DQF spectra with complementary properties will be sufficient for the determination of the complete geometry of small molecular (sub)units. The R^2-DQF sequence depicted in Fig. 2 [9] is particularly useful at R^2 conditions corresponding to high MAS frequencies and in the absence of chemical shieldings anisotropies. Other R^2-DQF schemes [13] are more suitable for spin systems characterized by large chemical shielding anisotropies [12]. By focussing on short-range order questions, this combined experimental / numerical R^2-DQF MAS NMR approach should be particularly useful in complementing diffraction experiments. Furthermore, it offers an experimental alternative for the indirect determination of molecular torsion angles from csa orientations for cases where a direct determination of these molecular geometries from so-called double-quantum heteronuclear local field experiments [14] is not possible, either due to the lack of a suitable $^1H,^{13}C$ spin-(sub)system or due to a lack of spatial isolation of the ^1H part of an otherwise suitable $^1H,^{13}C$ spin-(sub)system.

Acknowledgement

Support of our work by the Deutsche Forschungsgemeinschaft and the Fonds der Chemischen Industrie is gratefully acknowledged. X.H. acknowledges financial support by Aventis Pharma, Paris.

References

[1]   Andrew, E.R., Bradbury, A., Eades, R.G. and Wynn, V.T., Phys. Lett. 4 (1963) 99.

[2]   Raleigh, D. P., Levitt, M. H. and Griffin, R. G., Chem. Phys. Lett. 146 (1988) 71.

[3]   Levitt. M. H., Raleigh, D. P., Creuzet, F., and Griffin, R. G., J. Chem. Phys. 92 (1990) 6347.

[4]   Schmidt, A. and Vega, S., J. Chem. Phys. 96 (1992) 2655.

[5]   Nakai, T. and McDowell, C. A., J. Chem. Phys. 96 (1992) 3452.

[6]   Gregory, D. M., Wolfe, G. M., Jarvie, T. P., Sheils, J. C. and Drobny, G. P., Molec. Phys. 89 (1996) 1835.

[7]   Lee, Y. K., Kurur, N. D., Helmle, M., Johannessen, O. G., Nielsen, N. C. and Levitt, M. H., Chem. Phys. Lett. 242 (1995) 304.

[8]   For general review articles on recoupling methods under MAS NMR conditions see: i) Bennett, A. E., Griffin, R. G. and Vega, S., Recoupling of homo- and heteronuclear dipolar interactions in rotating solids, in: Solid-State NMR IV: Methods and Applications of Solid-State NMR, Vol. 33 NMR Basic principles and Progress (B. Blümich, Ed.), pp. 1-78, Springer Verlag, Berlin (1994); ii) Dusold, S. and Sebald A., Dipolar recoupling under magic-angle-spinning conditions, in: Annual Reports on NMR Spectroscopy, Vol. 41 (G. Webb, Ed.), pp. 185-264, Academic Press, London (2000); and references given therein.

[9]   Karlsson, T., Edén, M., Luthman, H. and Levitt, M. H., J. Magn. Reson. 145 (2000) 95.

[10]   Rach, W., Kiel, G. and Gattow, G., Z. Anorg. Allg. Chem. 563 (1988) 87.

[11]   Dusold, S. and Sebald, A., J. Magn. Reson. 145 (2000) 340.

[12]   Bechmann, M., Helluy, X. and Sebald, A., J. Magn. Reson. (2000), submitted.

[13]   Nielsen, N. C., Creuzet, F., Griffin, R. G. and Levitt, M. H., J. Chem. Phys. 96 (1992) 5668.

[14]   Feng, X., Lee, Y. K., Sandström, D., Edén, M., Maisel, H., Sebald, A. and Levitt. M. H., Chem. Phys. Lett. 257 (1996) 314.

Multiple-quantum spectroscopy of fully labeled polypeptides under MAS: A statistical and experimental analysis

Sorin Luca[a], Dmitri V. Filippov[b], Brigitta Angerstein[a], Gijs A. van der Marel[b], Jacques H. van Boom[b], and Marc Baldus[a]

[a] *Max-Planck-Institute for Biophysical Chemistry, Am Fassberg 11, 37077 Göttingen, Germany*
[b] *Leiden Institute of Chemistry, Leiden University, 2333 CC Leiden, The Netherlands*

Introduction

Spectral resolution represents a prerequisite for NMR based structural studies in multi- or fully labeled polypeptides such as membrane proteins, peptide ligands or protein aggregates. Recent applications in immobilized proteins [1-4] and a membrane protein aggregate [5] revealed that characteristic chemical shift information obtained in the liquid-state improves and expedites spectral assignment under MAS [6] conditions. In many cases, the observed NMR line width may limit applications in polypeptides and proteins of larger size. In this context, correlation experiments are desirable that maximize the spectral resolution without compromising the structural information contained in the spectra.

Multiple-quantum (MQ) spectroscopy [see e.g. 7,8] offers an additional degree of freedom since MQ precession frequencies, which must be detected in an indirect manner, are not only determined by the chemical shift of the involved resonances but also directly reflect the local molecular topology. Today, a variety of multi-pulse experiments are available to excite and reconvert MQ coherences under MAS conditions. These techniques have been used to obtain Double-quantum (DQ) - single-quantum (SQ) correlation spectra (usually termed INADEQUATE spectroscopy [9] in liquids) in a variety of solid phase compounds [10-16]. Obviously, the dispersion and correlation pattern depends on the spin system and the MQ order.

Under MAS and heteronuclear decoupling, MQ evolution in solids is mostly given by the chemical shift of the involved spins. Knowledge of the characteristic chemical shifts of natural amino acids hence permits prediction of MQ-SQ correlation results and their residue-specific spectral resolution. These spectra are presented here for a variety of MQ-SQ correlations and augmented by experimental results obtained in two uniformly

S.R.Kiihne and H.J.M.deGroot (eds.), Perspectives on Solid State NMR in Biology, 33-41.

[^{13}C,^{15}N]-labeled peptides. Our results demonstrate the spectral dispersion of 2D correlation spectra of variable coherence order. These relationships are useful for characterizing correlation spectra of immobilized polypeptides of unknown structure and may pave the way for additional NMR methods that retrieve structural information in the solid-state under a maximum degree of resolution.

Theory

The spectral resolution of MQ-SQ experiments is determined by the free precession periods of MQ and SQ coherences, respectively. In this context, products of single-spin operators are preferable to describe an arbitrary MQ coherence

$$|a\rangle\langle b| = \prod_{k=1}^{N} I_k^{\mu(k)} \tag{1}$$

where the $\mu(k)=+,-$ reflect changes in the magnetic quantum number $\Delta M_k = \pm 1$ of the active spins [8]. Passive spins, i.e. spins not directly involved in the formation of MQ coherence, occur with $\mu(k)=\alpha,\beta$ representing the spin states $|\pm 1/2\rangle$. MQ coherences are usually created during the preparation period [7,8,17,18] of a MQ correlation experiment (Fig. 1) using scalar or dipolar two-spin interactions. In the latter case, the excitation characteristics may strongly depend on the particular choice of pulse scheme and excitation time [19]. Isotropic scalar interactions, which have been successfully employed for through-bond correlation experiments in solids [20], could simplify the analysis but will be neglected in the discussion.

Fig. 1: General pulse scheme to observe MQ-SQ correlations during a two-dimensional NMR experiment (see e.g. [7,8]).

The Liouville –von Neumann equation,

$$\frac{d\sigma}{dt} = -i[H,\sigma], \tag{2}$$

that describes the time evolution of the density matrix σ can be used to calculate the precession frequencies during the t_1 MQ evolution. In spin 1/2 systems, the Hamiltonian H in eq. (2) usually contains time-independent components that represent the isotropic chemical shift terms and MAS-modulated contributions from anisotropic dipolar or chemical shift interactions. Standard commutation relations among single-spin operators can be used to evaluate eq. (2) for isotropic chemical shifts. Using

$$H_{iso} = \sum_k \Omega_k I_{kz}, \tag{3}$$

with Ω_k representing the chemical shift of spin k. We obtain the well known result [7,8] for the precession frequency during the t_1 evolution:

$$\omega_{ab} = \sum \Delta M_k \Omega_k \tag{4}$$

Additional contributions from the time-dependent parts of the system Hamiltonian can be neglected under fast MAS [21].

Destructive interference effects among the MQ transitions are minimized subsequently by employing reconversion pulse schemes that contain an effective dipolar (or scalar) Hamiltonian that is time-reversible [7,8]. In the context of broadband MQ spectroscopy in larger spin systems, MQ excitation and reconversion may, however, act on different spin subsystems. In the liquid state, this phenomenon is usually described by *direct* and *remote* peaks [see e.g. 22] and may lead to strong modulation of the observed MQ signal intensities. Similar observations were recently made in rotating solids [15] and may be used in the future for cross validation of the observed multiple-spin correlations. For the statistical predictions presented in the following, only the peak positions (ω_{ab}, Ω_l) and the dispersion as a function of amino acid type will be studied and compared to SQ-SQ (possibly MQ-filtered) correlation spectra.

Methods

The amino acids considered in this analysis and their statistical chemical shifts were obtained from the most recent BioMagRes data bank compilation [23]. For all 20 natural amino acids, ^{15}N and ^{13}C backbone and selected side chain chemical shift values were incorporated into a C++-based computer program to calculate MQ-SQ correlation spectra according to Eq. (4).

Experimental results were obtained on two uniformly [^{13}C, ^{15}N] labeled peptides. The pentapeptide RPYIL was prepared using standard FMOC solid phase chemistry with BOP/HOBt activation on an ABI 433A (Applied Biosystems/Perkin-Elmer) peptide synthesizer [24]. The peptide was purified with RP HPLC on an Alltima C18 column (10 x 250 mm, 5 μm particles, AllTech) applying a gradient of acetonitrile in 0.1 % aqueous TFA. The uniformly labeled trimer AGG was prepared by Boc-chemistry in solution starting from Boc-[U-^{13}C,^{15}N]-Gly-OMe. Stepwise elongation with Boc-[U-^{13}C,^{15}N]-Gly-OH and Boc-[U-^{13}C,^{15}N]-Ala-OH were accomplished by BOP/HOBt and EDC/HOBt methods, respectively. Semipermanent Boc protections were cleaved by 13 % HCl in EtOAc. The protected tripeptide AGG (Boc-Ala-Gly-Gly-OMe) was subsequently purified by silica gel column chromatography. Final deprotection was done by a consecutive treatment of AGG with 0.2 N NaOH (saponification of the methyl ester) and 95 % aqueous TFA (cleavage of the Boc-group) to furnish labeled AGG in 7 % overall yield. All isotope-labeled amino acid derivatives were purchased from Cambridge Isotope Laboratories, Inc. (USA).

Experimental NMR results were obtained on a 600 MHz WB instrument (Bruker, Karlsruhe) employing a 4mm triple channel MAS probe and low-temperature cooling unit that maintained a stable sample temperature of 277 K. An MAS frequency of 10 kHz was employed using TPPM decoupling [25] (using a r.f. field strength of 85 kHz) during evolution and detection periods. Phase sensitive MQ detection during the two-dimensional correlation experiments was accomplished using standard TPPI phase cycling [8].

Results and Discussion

In Fig.2, we begin with statistical predictions for an N-C polarization transfer experiment that correlates heteronuclear SQ coherences along the side chain of a polypeptide. Except for proline residues, these patterns correspond to $N-(C_\alpha, C_\beta, C_\gamma)$ - type connectivities in the aliphatic region. In agreement with experimental findings [4,5], characteristic chemical shift patterns are observed for G, T, S, P, A, and I residues. Note that the overall dispersion in the ^{15}N dimension amounts to about 15 ppm that significantly exceeds the resolution of a homonuclear ^{13}C-^{13}C experiment involving CO and C_α resonances. For this reason, heteronuclear SQ correlation spectroscopy [4,5] has proven to be essential for assignments of solid-phase (membrane) proteins under MAS. To further characterize the resolution obtainable from $N-(C_\alpha, C_\beta, C_\gamma)$ correlation spectra exemplified in Fig. 2, additional assumptions about the line width or the standard deviation (known for the liquid-state [22,23]) could be considered but will not be attempted here.

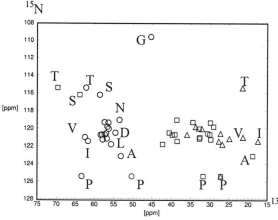

Fig. 2: Statistical prediction for heteronuclear $N-C_\alpha$ (circles), $N-C_\beta$ (squares) and $N-C_\gamma$ (triangles) (SQ,SQ) correlations. Amino acids with distinct correlations are indicated in single letter notation.

In many cases, homonuclear correlation experiments are to be preferred because of sensitivity reasons, labeling restrictions or the structural parameter under study. In Fig. 3, we compare two correlation experiments involving C_α backbone resonances during the detection of a two-dimensional MQ experiment. CO-C_α correlations detected in SQ mode are usually of limited use for spectral assignments because of the small chemical shift dispersion among carbonyl carbons. We hence concentrate in Fig. 3 on statistical predictions of a DQ-SQ (CO, C_α) experiment that are compared to statistical results of a triple quantum (TQ) – SQ experiment. In both cases, only the correlation pattern resulting from the C_α resonance (i.e. resulting from spin moieties CO-C_α and CO-C_α-C_β in the DQ and TQ case, respectively) in t_2 is shown. Comparison of both correlation patterns immediately reveals an increased resolution for the TQ-SQ approach. The resolution in the DQ dimension is comparable to results of the heteronuclear SQ-SQ correlation pattern in Fig. 2. We also note that similar results were obtained in a ZQ (zero-quantum)-SQ model correlation study (data not shown). The inspection of Fig. 3 shows that, in addition to Fig. 2, the TQ-SQ experiment may reveal L, D, N, A and, possibly F and Y residues with reasonable resolution. Combining information of both experiments hence could lead to the identification of 12 of the 20 amino acids, provided that overall line width does not exceed the observed dispersion.

Further information can be gained by including the TQ correlation patterns observed at the C_β resonance and are shown in Fig. 4. Depending on the primary sequence of the polypeptide under study, additional correlations involving C, H, or K, M residues (indicated in Fig. 4) could be detected. Since a MQ-SQ correlation experiment reconverts MQ coherence to SQ signals for all spins involved in the formation of a particular MQ coherence, horizontal lines can be used for cross validation. Preliminary 2D TQ-SQ experiments in model peptides [26] are in good agreement with these statistical predictions. Obviously, the presented results do not take

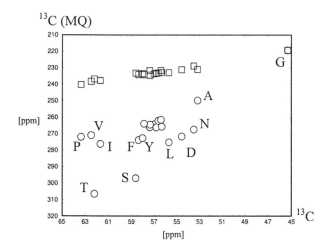

Fig.3. Comparison of statistical predictions for a DQ-SQ (squares) and a TQ-SQ (circles) experiment. Correlation patterns for C_α detected DQ and TQ coherence involving CO, C_α and CO,C_α and C_β resonances, respectively, are shown.

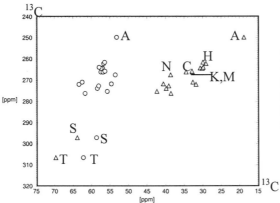

Fig.4. Statistical predictions for a TQ-SQ experiment. Correlation patterns for C_α (circles) and C_β (triangles) detected TQ coherence involving CO, C_α and C_β resonances are shown.

into account sensitivity considerations or changes in line width during MQ evolution periods. On the other hand, significant progress has been made in the design of ZQ and DQ excitation schemes that are based on a variety of dipolar recoupling methods (see e.g. [27]) and can be modified according to standard procedures [7,8] for MQ spectroscopy. DQ excitation and filtering has been demonstrated using a variety of approaches. Experimental TQ filtering efficiencies around 8 % of the direct CP signal were reported recently in U-[^{13}C] labeled L-alanine under MAS [28]. Antzutkin and Tycko observed 10-quantum ^{13}C-^{13}C- coherences under static conditions [29]. Experimental modifications that enhance particular MQ signals (such as discussed in Ref. [28]) could improve the possibilities of profiting from the resolution enhancement suggested by the results of Figs. 3 and 4.

In the current context, we restrict ourselves to an experimental verification of the predictions of DQ-SQ spectroscopy on two uniformly ^{13}C and ^{15}N labeled model peptides, AGG and RPYIL. For both peptides, we focus on the aliphatic region that

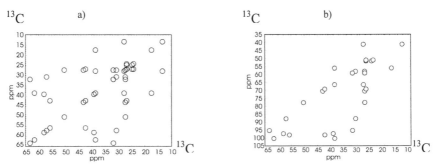

Fig. 5. Statistical predictions for a SQ-SQ (a) and DQ-SQ (b) experiment on RPYIL.

usually contains the largest spectral dispersion. Fig. 5 contains a statistical prediction of the SQ-SQ (a) and DQ-SQ (b) correlation patterns for the 5-residue peptide RPYIL. Note the reduced number of peaks in the latter experiment and the increased resolution along t_1.

In the case of AGG, we expect a single DQ-SQ correlation in the side chain region that can be used as a reference for signal to noise and for line width considerations. Experimental results are displayed in Fig. 6. For excitation and reconversion, an amplitude-modulated version of the HORROR [30] recoupling scheme was used [31-33] with a maximum r.f. field strength during recoupling of 5 kHz. Besides advantages in the orientational dependence of the excitation [30] and a large effective dipolar coupling [30,27], this method avoids significant MQ sideband intensities in the MQ dimension [19]. Results for the penta-peptide RPYIL are shown in Fig. 6b) using the same spectral window as in the predictions, above.

In spite of the larger line width observed in the experimental spectra, the overall pattern predicted in Fig. 5 is well reproduced. Small deviations are found for correlations involving the proline residue that could possibly result from the particular geometry of the $C_{\alpha-\delta}$ resonances in this residue. The relatively large SQ line width observed in both samples is most likely caused by inhomogeneous broadening, possibly due to lyophilization. For both peptides, the largest line width is observed for C_α and C_β resonances. These observations would be consistent with liquid-state NMR studies that have documented a direct influence of the dihedral angles upon the resonance line position in polypeptides [see e.g. 34,35]. The results obtained on AGG can be used to compare the effective MQ and SQ line width. In agreement with recent relaxation studies [36], we observe for both peptides an increased line width for the DQ dimension that reduces the beneficial effect of MQ evolution dimensions

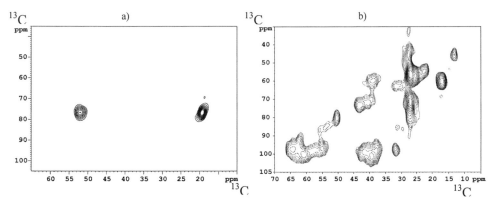

Fig. 6. Experimental DQ-SQ spectra on a) AGG and b) RPYIL.

Conclusions

MQ spectroscopy represents an established tool in liquid-state NMR for spectral simplification and resolution enhancement involving proton resonances. Our statistical analysis shows that significant improvements can also be expected for solid-state MAS NMR experiments of multi- or fully labeled polypeptides. Experimental results in two model peptides indicate that changes in the relaxation behavior may partially compensate for these improvements. In addition, experimental realizations of MQ correlation experiments will benefit from improvements in MQ excitation efficiency or from more sophisticated methods that e.g. involve proton evolution or detection periods.

The presented study gives a general compilation of MQ spectra in solid-phase polypeptides and may be helpful in the design and application of correlation experiments under MAS.

References

[1]   Straus, S.K., Bremi, T.and Ernst, R.R., J. Biomol. NMR 12 (1998) 39.
[2]   Hong, M. J. Biomol. NMR 15 (1999) 1.
[3]   McDermott, A., Polenova, T., Bockmann, A., Zilm, K.W., Paulsen, E.W., Martin, R.W. and Montelione, G.T. J. Biomol. NMR 16 (2000) 209.
[4]   Pauli, J., Baldus, M., van Rossum, B., de Groot, H. and Oschkinat, H, Chem.BioChem (2001) in press.
[5]   Egorova-Zachernyuk, T.A., Hollander, J., Fraser, N., Gast, P., Hoff, A.J., Cogdell, R., de Groot, H..J.M. and Baldus, M. J. Biomol. NMR (2001) in press.
[6]   Andrew, E.R., Bradbury, A. and Eades, R.G. Nature 182 (1958) 1659.
[7]   Yen, Y.S. and Pines, A., J.Chem.Phys. 78 (1983) 3579.
[8]   Ernst, R.R , Bodenhausen, G. and Wokaun, A. (1987) Principles of Nuclear Magnetic Resonance in One and Two Dimension, Oxford: Claredon Press. Chapter 5.
[9]   Bax, A. Freeman, R. and Kempsell, S.P., J.Amer.Chem.Soc. 55 (1980) 4849.
[10]   Menger, E.M., Vega, S.and Griffin, R.G., J.Amer.Chem.Soc. 108 (1986) 2215.
[11]   Geen, H., Titman, J.J., Gottwald, J. and Spiess, H.W., Chem.Phys.Lett. 227 (1994) 79.
[12]   Feike, M., Graf, R., Schnell, I., Jager, C. and Spiess, H.W., J.Amer.Chem.Soc. 118 (1996) 9631.
[13]   Dollase, W.A., Feike, M., Foerster, H., Schaller, T., Schnell, A., Sebald, A. and Steuernagel, J.Amer.Chem.Soc. 119 (1997) 3807.
[14]   Lesage, A., Auger, C., Caldarelli, S. and Emsley, L., J.Amer.Chem.Soc. 119 (1997) 7867.
[15]   Howhy, M., Rienstra, C.M., Jaroniec, C.P. and Griffin, R.G., J.Chem.Phys. 110 (1999) 7983.
[16]   Hong, M., J.Mag.Res. 136 (1999) 86.
[17]   Baum, J., Munowitz, M., Garroway, A.N. and Pines, A., J.Chem.Phys. 83 (1985) 2015.
[18]   Munowitz, M., Pines, A., and Mehring, M., J.Chem.Phys. 86 (1987) 3172.
[19]   Geen, H., Titman, J.J., Gottwald, J. and Spiess, H.W., J.Mag.Res. A 114 (1995) 264.
[20]   Baldus, M. and Meier, B.H., J. Mag. Res. A 121 (1996) 65.
[21]   Goldman, M., J.Mag.Res. A 102 (1993) 173.
[22]   Cavanagh, J., Fairbrother, W.J., Palmer, A.G., Skelton, N.J., (1995) *Protein NMR Spectroscopy Principles and Practice*, Academic Press.

[23] BioMagRes data bank, A Repository for Data from NMR Spectroscopy on Proteins, Peptides, and Nucleic Acids, see also: http://www.bmrb.wisc.edu/pages/

[24] For a review about Fmoc-chemistry see *e.g.*: Atherthon, E. & Sheppard, R. C.*Solid Phase Peptide Synthesis: A Practical Approach,* IRL Press: Oxford, 1989.

[25] Bennett, A.E., Rienstra, C.M., Auger, M., Lakshmi, K.V., and Griffin, R.G. J.Chem.Phys. 103 (1995) 6951.

[26] Luca, S., Angerstein, B., and Baldus, M., manuscript in preparation.

[27] Baldus, M., Geurts, D.G. and Meier, B.H. Solid State NMR, 11 (1998) 157.

[28] Eden, M. and Levitt, M.H., Chem.Phys.Lett. 293 (1998) 173.

[29] Antzutkin, O.N. and Tycko, R., J.Chem.Phys. 110 (1999) 2749.

[30] Nielsen, N.C., Bildsoe, H., Jakobsen, H.J. and Levitt, M.H., J.Chem.Phys. 101 (1994) 1805

[31] Baldus, M., van Os, J. and Meier, B.H., proceedings of the 38th ENC conference, Orlando/USA, (1997), p. 152.

[32] Verel, R., Baldus, M., Ernst, M. and Meier, B.H., Chem. Phys. Letters, 287 (1998).

[33] Baldus, M., Rovnyak, D. and Griffin, R.G., J.Chem.Phys., 112, (2000) 5902.

[34] Wishart, D.S. and Sykes, B.D., J.Biomol. NMR 4 (1994) 171.

[35] De Dios, A.C., Pearson, J.G. and Oldfield, E., Science 260 (1993) 1491.

[36] Karlsson, T., Brinkmann, A., Verdegem, P.J.E., Lugtenburg, J., and Levitt, M.H., Solid State NMR, 14 (1999) 43.

SECTION II:

Advances in Techniques for Aligned Systems

^{2}H, ^{15}N and ^{31}P solid-state NMR spectroscopy of polypeptides reconstituted into oriented phospholipid membranes

Burkhard Bechinger, Christopher Aisenbrey, Christina Sizun, and Ulrike Harzer

Max Planck Institute for Biochemistry, 82152 Martinsried, Germany

Introduction

Although membrane proteins are abundant and fulfill many important functions, only a few high-resolution conformations of this class of proteins have been published (reviewed in [1; 2]). Structural investigations are hampered by the problems that are encountered during their large-scale expression and purification as well as during application of diffraction and NMR techniques. Since publication of the high-resolution diffraction map of the photoreaction centre [3], the conformations of other bacterial membrane proteins have been solved at a relatively slow pace. Most of these proteins already occur at high concentrations in nature and some of them even in ordered arrays [1; 2]. More recently high resolution structures of eukaryotic membrane proteins have been added to the structural data base, including rhodopsin, a G protein-coupled receptor [4; 5]. Improvements in crystallisation as well as in X-ray and electron diffraction techniques hold promise that these developments will accelerate, and structures of an increasing number of membrane proteins will become available in the future. Some of this work illuminates structural changes that occur during functional activities in a stroboscopic manner [4; 6].

Proteins that exhibit a more flexible character, however, are not fully accessible by diffraction techniques, and this is also where NMR spectroscopy can provide interesting structural insights in a unique manner. For example, the α-helix of mellitin is considerably shorter when associated with lipid bilayers [7; 8] than in the presence of detergent micelles [9; 10], in methanol [11; 12] or in the crystal [13]. In a related manner, biochemical and biophysical investigations show that the accessibility of residues of the hydrophobic helices of the *E. coli* protein colicin Ia change with experimental conditions [14] suggesting a more dynamic model in which topological and/or conformational alterations occur. A similar model might also apply to colicin E1

S.R.Kühne and H.J.M.deGroot (eds.), Perspectives on Solid State NMR in Biology, 45-53.

where the average penetration depth of tryptophan residues is altered as a function of bilayer thickness [15]. Therefore, similar to observations made for smaller polypeptides, several equilibria can describe the different topologies of colicins: soluble ⇔ surface associated ⇔ inserted, 'pen-knife' model ⇔ inserted, 'umbrella' model ⇔ etc. [16].

Two fundamentally different approaches are used in solid-state NMR spectroscopy to investigate selected questions in a highly accurate manner. First, fast averaging of magnetic interactions by magic angle spinning NMR results in spectra representing the isotropic chemical shifts as well as spinning side band intensities. By rotating the sample around the magic angle, valuable structural information is lost. This information is selectively reintroduced by manipulating the spins in a well-defined manner [17]. Although MAS spectra resemble those obtained in solution, the line widths obtained by MAS-NMR remain comparatively broad.

Second, static samples exhibit the residual anisotropies of chemical shifts, dipolar and quadrupolar interactions. In non-oriented samples (powder patterns) these can be analysed with regard to local and global motions [18]. When the samples are aligned with respect to the magnetic field direction, however, valuable structural information is obtained from orientational constraints [19]. More recently, oriented samples have also been introduced into magic angle spinning investigations by aligning the sample material on stacks of round glass plates with diameters that exactly match the inside inside diameter of 7 mm MAS rotors [20]. Strong adhesion of the membranes onto the glass plates allows for slow to moderate sample spinning speeds. In these preparations the normal of the membranes is oriented parallel to the magic angle spinning axis, therefore, rotational diffusion of the molecules around this axis efficiently averages the anisotropies of NMR interactions [20]. Whereas membrane undulations of vesicular membranes occur on slow time scales and interfere with averaging by magic angle sample spinning (MAS). These fluctuations are suppressed by mechanically supporting lipid bilayers. The resonance line-widths in magic angle oriented sample spinning (MAOSS) experiments, therefore, are superior even when compared to fast MAS solid-state NMR spectra of non-oriented membrane samples. Topological information is extracted from the MAOSS spectra by quantitatively analysing the spinning side band intensities (Figs. 1E-F). The spinning speeds accessible with this set-up are, however, limited due to the mechanical stability of the samples and the centrifugal forces that expel the oriented membranes from in between the mechanical supports. Although uniformly labelled polypeptides have been investigated by oriented solid-state NMR spectroscopy (e.g.[21]), static and magic angle spinning solid-state NMR approaches require the further development of assignment experiments in order to extract the full range of structural information that is inherently present in solid-state NMR spectra (e.g. [22-26]). In addition, the line widths obtained by solid-state NMR spectroscopy are broader than those observed in solution. The consequently lower signal-to-noise ratio and decreased resolution impose severe restrictions on the experimentalist [27]. Therefore, in order to obtain detailed structural information and to unambiguously assign individual residues, both solid-state NMR approaches have so far heavily relied on the labelling of residues with stable isotopes (e.g. ^{15}N, ^{13}C, ^{2}H) at a single or a few selected sites.

Methods

Peptides were prepared on ABI 430 or Millipore 9050 automated peptide synthesizers using Fmoc protected amino acids [28; 29]. Amino acids labelled with stable isotopes were obtained from Campro (Emerich, Germany) or Promochem (Wesel, Germany). High purity of the final products was ensured by reversed-phase preparative and analytical HPLC using acetonitril / water gradients (0.1 % TFA) and mass spectrometric analysis.

After the pH of the peptides had been adjusted, homogenous solutions in organic solvents of the peptides and phospholipids (Avanti Polar Lipids, Birmingham, AL) were prepared and applied onto clean ultra thin cover glasses (9 x 22 mm). After equilibration at defined relative humidity (usually 93 %), the glass plates were stacked on top of each other (Fig. 1A) and the samples carefully sealed [30]. Alternatively, oriented membranes were prepared on long sheets of thin polymers such as polyetheretherketone (600 x 10 mm, Goodfellow, Huntingdon, England). Once the membrane samples have been equilibrated, the membrane stacks / polymer sheets are wrapped into a continuous spiral with an outer diameter that fits into the MAS rotor [31; 32]. MAS spectra were simulated on a Silicon Graphics Indigo II workstation using a Fortran alghorithm based on the fomalism published in reference [33].

Proton-decoupled ^{15}N and ^{31}P solid-state NMR spectra were obtained on a Bruker Avance solid-state NMR spectrometer operating at 9.4 Tesla using cross polarization [34; 35] and Hahn echo pulse sequences [36], as described previously [37-39]. Alternatively, proton-decoupled ^{31}P MAS NMR spectra were obtained using a single ^{31}P pulse (3.5 us), recycle delays of 5 s, and a 6.45 ms acquisition time. Chemical shifts are referenced to ammonium chloride (41.5 ppm) or phosphoric acid. Deuterium NMR spectra were recorded using a quadrupolar echo pulse sequence [40].

Results and Discussions

The orientation-dependent ^{15}N chemical shift as obtained from proton-decoupled ^{15}N solid-state NMR spectroscopy provides a direct indicator for the approximate alignment of membrane-associated α-helices with respect to the bilayer normal [30; 32; 41]. This approach is based on reconstitution of polypeptides into phospholipid membranes oriented with their bilayer normal preferentially parallel to the magnetic field direction (Fig. 1A). By measuring a single ^{15}N chemical shift resonance the helix tilt angle can be estimated. Whereas chemical shifts (σ) < 100 ppm indicate alignments in-plane with the membrane surface, transmembrane orientations exhibit resonances around 220 ppm. By measuring the chemical shifts of several residues in the membrane spanning helices of Vpu of HIV-1 and M2 of Influenza virus A, their tilt angles were determined to be around 20 and 35 degrees, respectively [30; 42; 43]. Similarly, the in-plane orientation

B. BECHINGER, ET AL.

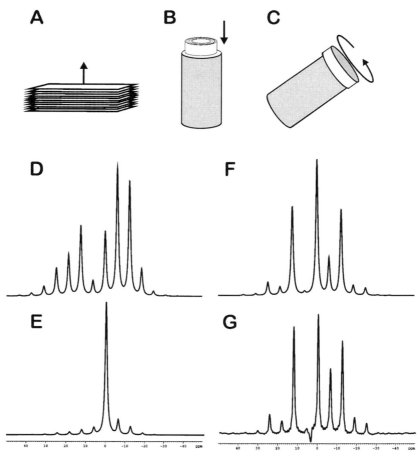

Figure 1: Examples of geometrical arrangements of oriented membranes used for solid-state NMR investigations. A) Stacks of cover glasses (typically 9x 22 mm) with approximately 3000 lipid bilayers in between each pair. The arrow indicates the direction of the membrane normal. B) Long sheets of polymer covered with oriented membranes are rolled into a spiral. Thereafter, these samples are inserted into the MAS rotor with the axis of the membrane normal perpendicular to the spinner axis. C) This arrangement allows for fast spinning of oriented membranes at the magic angle. Calculated proton-decoupled ^{31}P NMR spectra of liquid crystalline phosphatidylcholine membranes are shown: D) Powder pattern, E) lipid long axes at orientations parallel to the rotor axis, and F) perpendicular to the rotor axis as obtained by the geometrical arrangement shown in (B,C). G) Experimental proton-decoupled single-pulse ^{31}P MAS spectrum of POPC bilayers oriented on polymer sheets. A similar situation has been simulated in panel (F). The following parameters were used for the calculations: Chemical shift anisotropy $\delta_{||}-\delta_{\perp} = 45$ ppm, MAS frequency 993 Hz, Lorentzian line broadening 100 Hz and mosaic spread $4°$. A contribution of 20 % non-oriented lipid was included in the simulations shown in panels (E, F).

of the first cytoplasmic helix of Vpu was derived from ^{15}N chemical shift information and shows excellent agreement with a hydrophobic moment analysis [44]. Although this helix seems to be associated less tightly with the membrane when two neighbouring serines (at positions 52 and 56) are phosphorylated, the alignment of the membrane-associated helices remains largely unaltered [44].

The ^{15}N chemical shift and the [1]H-^{15}N dipolar coupling alone provide the approximate helix orientation in a direct manner due to the close to parallel alignment of the unique chemical shift tensor element σ_{33} and the [1]H-^{15}N vector, respectively, with respect to the helix axis [30; 45]. Investigations of ^{13}C labelled carbonyls of peptide bonds have also been used to extract a transmembrane orientation of mellitin in oriented bilayers [46]. Consideration of additional constraints enables a more accurate determination of helix orientation. For example, separated local field experiments allow the simultaneous extraction of both the ^{15}N chemical shift as well the [1]H-^{15}N dipolar coupling. By determining these parameters from consecutive residues of bilayer-associated magainin 2, a right-handed α-helical conformation oriented parallel to the membrane surface has been selected among other low-energy conformations [41; 47]. Interestingly this polypeptide has been shown to exhibit characteristics of single channels in some electrophysiological experiments (discussed in [16]). In a related manner, the voltage-sensing peptide S4 of the rat brain sodium channel results in several conductance levels across phospholipid bilayers [48]. One of them is continuously open and, therefore, correlates with the low energy conformations tested by structural studies. By combining multidimensional high-resolution NMR experiments on S4 in DPC micelles, quenching by DOXYL paramagnetic spin labels, oriented ^{15}N solid-state NMR spectroscopy and molecular modelling, a consistent view of the peptide oriented in the plane of the lipid bilayer is obtained [28]. Furthermore, Cross and co-workers have obtained 120 orientational constraints for gramicidin A reconstituted into oriented DMPC membranes by combining ^{15}N and ^{13}C chemical shift, dipolar coupling as well as ^{2}H quadrupolar splitting information [19]. These data have been used to calculate the backbone and side chain high-resolution structure of this peptide.

In order to accurately translate measured values of the chemical shift and the dipolar interactions into high-resolution structures of membrane polypeptides, a detailed description of the orientation and size of the chemical shift tensor elements is required for every individual position to be analysed (reviewed in [19; 41]). Whereas the chemical shifts and dipolar interactions are measured accurately, major uncertainties in the analysis of oriented spectra arise from variance of the tensor elements [18; 42; 49]. In this respect the investigation of single-site labelled polypeptides has distinct advantages over uniformly labelled proteins as the main tensor elements of the former samples can be analysed in considerable detail. Investigations of powder pattern samples of single site labels thereby reveal dynamic averaging as well as side chain-, hydration- and conformation-dependent variations of the NMR interactions with the magnetic field [18].

The ^{15}N chemical shift information has been used to investigate the interaction contributions that are important for the alignment of α-helical peptides. Using histidine-containing peptides, pH-sensitive molecular switches have been designed [37]. Our experiments show that during the transfer of residues from the aqueous phase (or bilayer interface) to the membrane interior, the energy required to discharge amino acid side chains as well as hydrophobic and polar interactions is important [37; 37; 50]. These histidine-containing peptides exhibit a similar or improved degree of antibiotic activity when compared to their natural analogues, in particular when investigated at pH values where they align along the surface of the membrane [51]. More recently, the energy contributions of lysine side chains have been investigated in more detail [39]. Whereas the unfavourable energy contributions associated with placement of one lysine in the bilayer interior are compensated by the sum of hydrophobic interactions of a transmembrane sequence, the presence of three or more lysines results in in-plane alignments of amphipathic helical peptides. Finally, the effects of hydrophobic mismatch on a series of polypeptides reconstituted in phospholipid membranes of varying hydrophobic thicknesses have been studied using mechanically supported membrane samples [38]. Whereas a high degree of mutual adaptation exists when the peptides are a little shorter or much longer than the hydrophobic thickness of the membrane, the orientational order of the sample is lost when the amount of hydrophobic mismatch between the phospholipid bilayer and the peptides further increases. Our ^{15}N chemical shift measurements of aligned transmembrane model peptides also provide a first experimental indication of changes in helix tilt angle to compensate for the hydrophobic mismatch of long peptides. In addition, solid-state NMR investigations reveal that peptides that are too short to span the membrane interact with the phospholipid bilayers in a different manner when compared to those that exceed the membrane hydrophobic thickness.

While the ^{15}N chemical shift of individual residues provides direct access to the investigation of helix tilt angles with respect to the membrane normal, this parameter is only sensitive to rotation around the helix axis in exceptional cases [44]. Therefore, either information from a multitude of ^{15}N labelled sites, indirect arguments, or additional constraints from additional isotopic labels are required [19; 42; 43]. The ^{2}H quadrupolar splitting of ^{2}H$_3$ labelled alanine and the ^{15}N chemical shift of backbone labelled residues provide complementary information about polypeptide alignments (Fig. 2). Fast rotation around the C_α-C_β axis of alanine results in the observation of an averaged quadrupolar splitting for three deuterons of the methyl group. The quadrupolar splitting of the C-^{2}H$_3$ group, therefore, also provides a direct measure of the polypeptide backbone alignment. Whereas the ^{15}N measurement is mostly sensitive to changes in tilt angle (Fig. 2A), the quadrupolar splitting provides an additional indicator of rotation around the helix long axis (Fig. 2B). We have succeeded in obtaining oriented ^{2}H NMR spectra of ^{2}H$_3$-Ala-model peptides reconstituted in oriented phospholipid membranes. These data are currently being analysed in more detail to investigate the effects of motional averaging.

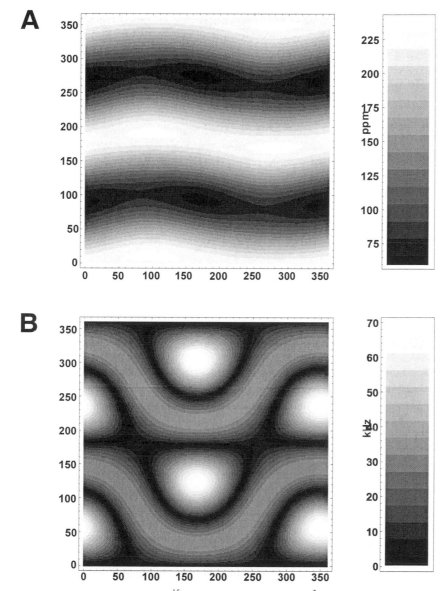

Figure 2: Contour plot showing A.) ^{15}N chemical shifts and B) ^{2}H quadrupolar splittings as a function of rotation around two axes perpendicular to each other. These calculations are obtained from a peptide reconstituted into oriented phospholipid bilayers and labelled at a single alanine residue either with $^{2}H_3$-methyl-alanine or with backbone ^{15}N. Fast rotation around the C_α-C_β bond results in a single quadrupole splitting of all three sites. Due to the complementary nature of the ^{2}H and ^{15}N measurements very different patterns are obtained.

More recently, oriented membranes have been used in magic angle spinning experiments [20]. This approach allows one to obtain high resolution MAS spectra and at the same time to derive orientational information from spinning side bands (Figs. 1D-G). In order to facilitate sample preparation and at the same time increase the range of applicable spinning speeds, we have applied oriented membranes onto thin polymer films (Fig. 1B). In contrast to previous MAOSS approaches, the membranes on the polymer sheets are oriented with their normal parallel to the direction of the centrifugal forces (Figs. 1B,C). Therefore, the magic angle spinning experiment can be performed at low or high spinning speeds. In this new approach the degree of sample alignment can improve by spinning the sample [31; 52]. Orientational information is obtained by analysing the side band intensities at low spinning speeds as illustrated in Figures 1D-G [20; 31]. Multidimensional high spinning speed experiments with isotropic resolution in one dimension and interference with MAS averaging by appropriate pulse sequences in a second dimension can be envisioned using this approach (e.g. [53] and references cited therein). In addition, using this kind of oriented MAS sample, measurements of orientation-dependent dipolar interactions containing information about molecular alignment as well as interatomic distances are possible.

References

[1]   Garavito, R.M. (1998) *Curr.Opin.Struct.Biol. 9*, 344-349
[2]   Tsukihara, T. & Lee, S.J. (1999) *J.Synchrotron Rad. 6*, 918-927
[3]   Deisenhofer, J., Epp, O., Miki, K., Huber, R., & Michel, H. (1985) *Nature 318*, 618-618
[4]   Toyoshima, C., Nakasako, M., Nomura, H., & Ogawa, H. (2000) *Nature 405*, 647-654
[5]   Palczewski, K., Kumasaka, T., Hori, T., Behnke, C.A., Motoshima, H., Fox, B.A., Le Trong, I., Teller, D.C., Okada, T., Stenkamp, R.E., Yamamoto, M., & Miyano, M. (2000) *Science 289*, 739-745
[6]   Kühlbrandt, W. (2000) *Nature 406*, 569-570
[7]   Dempsey, C.E. & Butler, G.S. (1992) *Biochemistry 31*, 11973-11977
[8]   Okada, A., Wakamatsu, K., Miyazawa, T., & Higashijima, T. (1994) *Biochemistry 33*, 9438-9446
[9]   Brown, L.R., Braun, W., Kumar, A., & Wüthrich, K. (1982) *Biophys.J. 37*, 319-328
[10]   Ikura, T., Go, N., & Inagaki, F. (1991) *Proteins 9*, 81-89
[11]   Bazzo, R., Tappin, M.J., Pastore, A., Harvey, T.S., Carver, J.A., & Campbell, I.D. (1988) *Eur.J.Biochem. 173*, 139-146
[12]   Dempsey, C.E. (1988) *Biochemistry 27*, 6893-6901
[13]   Terwilliger, T.C., Weissman, L., & Eisenberg, D. (1982) *Biophys.J. 37*, 353-361
[14]   Kienker, P.K., Qiu, X., Slatin, S.L., Finkelstein, A., & Jakes, K.S. (1997) *J.Membrane Biol. 157*, 27-37
[15]   Malenbaum, S.E., Merrill, A.R., & London, E. (1998) *J.Nat.Toxins. 7*, 269-290
[16]   Bechinger, B. (2000) *Phys.Chem.Chem.Phys. 2*, 4569-4573
[17]   McDowell, L.M. & Schaefer, J. (1996) *Curr.Opin.Struct.Biolo. 6*, 624-629
[18]   Lazo, N.D., Hu, W., & Cross, T.A. (1995) *J.Magn.Res. 107*, 43-50
[19]   Cross, T.A. (1997) *Methods Enzymol. 289*, 672-696
[20]   Glaubitz, C. & Watts, A. (1998) *J.Magn.Reson. 130*, 305-316

[21] Lambotte, S., Jasperse, P., & Bechinger, B. (1998) *Biochemistry 37*, 16-22
[22] Straus, S.K., Bremi, T., & Ernst, R.R. (1998) *J.Biomolec.NMR 12*, 39-50
[23] Hong, M. & Griffin, R.G. (1998) *Journal of the American Chemical Society 120*, 7113-7114
[24] Hong, M. (1999) *J.Biomol.NMR 5*, 1-14
[25] Tian, F., Fu, R., & Cross, T.A. (1999) *J.Magn.Reson. 139*, 377-381
[26] Baldus, M. & Meier, B.H. (1996) *J.Magn.Res. 121*, 65-69
[27] Tycko, R. (1996) *J.Biomol.NMR 8*, 239-251
[28] Mattila, K., Kinder, R., & Bechinger, B. (1999) *Biophys.J. 77*, 2102-2113
[29] Henklein, P., Schubert, U., Kunert, O., Klabunde, S., Wray, V., Kloppel, K.D., Kiess, M., Portsmann, T., & Schomburg, D. (1993) *Peptide Research 6*, 79-87
[30] Bechinger, B., Kinder, R., Helmle, M., Vogt, T.B., Harzer, U., & Schinzel, S. (1999) *Biopolymers 51*, 174-190
[31] Kinder, R. (1999) *PhD, Technical University Munich*
[32] Bechinger, B. (2000) *Molec.Membr.Biol.(in press)*
[33] Mehring, M. (1983) in *Principles of High Resolution NMR in Solids*, Springer, Berlin.
[34] Pines, A., Gibby, M.G., & Waugh, J.S. (1973) *J.Chem.Phys. 59*, 569-590
[35] Levitt, M.H., Suter, D., & Ernst, R.R. (1986) *J.Chem.Phys. 84*, 4243-4255
[36] Rance, M. & Byrd, R.A. (1983) *J.Magn.Res. 52*, 221-240
[37] Bechinger, B. (1996) *J.Mol.Biol. 263*, 768-775
[38] Harzer, U. & Bechinger, B. (2000) *Biochemistry, in press*
[39] Vogt, T.C.B., Ducarme, P., Schinzel, S., Brasseur, R., & Bechinger, B. (2000) *Biophys.J. 79*, 4644-4656
[40] Davis, J.H., Jeffrey, K.R., Bloom, M., Valic, M.I., & Higgs, T.P. (1976) *Chem.Phys.Lett. 42*, 390-394
[41] Bechinger, B. (1999) *Biochim.Biophys.Acta 1462*, 157-183
[42] Wray, V., Kinder, R., Federau, T., Henklein, P., Bechinger, B., & Schubert, U. (1999) *Biochemistry 38*, 5272-5282
[43] Kovacs, F.A. & Cross, T.A. (1997) *Biophys.J. 73*, 2511-2517
[44] Henklein, P., Kinder, R., Schubert, U., & Bechinger, B. (2000) *FEBS Lett.(in press)*
[45] Bechinger, B., Gierasch, L.M., Montal, M., Zasloff, M., & Opella, S.J. (1996) *Solid State NMR 7*, 185-192
[46] Smith, R., Separovic, F., Milne, T.J., Whittaker, A., Bennett, F.M., Cornell, B.A., & Makriyannis, A. (1994) *J.Mol.Biol. 241*, 456-466
[47] Bechinger, B., Zasloff, M., & Opella, S.J. (1993) *Protein Sci. 2*, 2077-2084
[48] Tosteson, M.T., Auld, D.S., & Tosteson, D.C. (1989) *P.N.A.S.USA 86*, 707-710
[49] Teng, Q. & Cross, T.A. (1989) *J.Magn.Reson. 85*, 439-447
[50] Kinder, R. & Bechinger, B. (1999) *Biophys.J. 76*, A443-
[51] Vogt, T.C.B. & Bechinger, B. (1999) *J.Biol.Chem. 274*, 29115-29121
[52] Gröbner, G., Taylor, A., Williamson, P.T.F., Choi, G., Glaubitz, C., Watts, J.A., deGrip, W.J., & Watts, A. (1997) *Analyt.Biochem. 254*, 132-138
[53] Gross, J.D., Costa, P.R., & Griffin, R.G. (2000) *J.Chem.Phys. 108*, 7286-7293

From Topology to High Resolution Membrane Protein Structures

T.A. Cross,[a,c,d] **S. Kim,**[c,d] **J. Wang,**[c,d] **& J.R. Quine,**[b,d]
*[a]Department of Chemistry, [b]Department of Mathematics, [c]Institute of Molecular Biophysics &
[d]National High Magnetic Field Laboratory,
Florida State University, Tallahassee, FL 32310 USA*

Introduction

Structural genomics of membrane proteins represents a particularly exciting challenge for biophysicists. These proteins are heterogeneous, both in themselves because of post translational modifications and in their environment from the heterogeneity of the membrane. Such heterogeneity greatly complicates structural characterization. Furthermore, the scarcity of specific interactions between secondary structural elements in the membrane environment results in increased dynamics and flexibility. Solid state NMR provides a methodology and a range of techniques that can characterize these proteins in a planar lipid bilayer environment [1]. NMR is an inherently high resolution method in that specific atomic sites are directly observed. With solid state NMR this potential for high resolution is fulfilled by obtaining very precise distance and orientational restraints. Until recently however, topology was very difficult to achieve; now several approaches are evolving so that both low resolution topology and unique high resolution structures can be obtained.

As an initial step towards functional understanding of the proteins resulting from an entire genome, the aim of protein structural genomics is to gain a rapid structural characterization for a large fraction of a proteome. While a variety of biophysical approaches from X-ray crystallography to electron diffraction, solution and solid state NMR will have a role in structural characterization, the step beyond structure requires a detailed characterization of dynamics and chemical properties to understand function. Such characterizations are at the heart of the strengths offered by NMR spectroscopy. While order parameter characterizations are popular today, far more detailed descriptions of molecular motions can, and have been achieved through characterizations of powder pattern averaging coupled with relaxation analysis [2].

S.R.Kiihne and H.J.M.deGroot (eds.), Perspectives on Solid State NMR in Biology, 55-69.
©2001 Kluwer Academic Publishers. Printed in the Netherlands.

Chemical properties such as H/D exchange, pK_a values and metal ion binding can be exquisitely characterized for specific sites in a protein while it is in a native-like environment. This integrated view of structure, dynamics and chemistry from NMR will lead to a detailed understanding of membrane protein function and an achievement of the ultimate structural genomics goal.

Membrane proteins comprise 30 - 40% of known proteomes and yet remarkably few of their structures have been solved to atomic resolution. In fact, less than 1% of the structures in the Protein Data Bank are membrane protein structures. While difficulties in forming 2-dimensional and 3-dimensional crystals have contributed to this scarcity of structure, it has also been very difficult to produce adequate quantities of membrane protein at the high purity required for structural characterization. Higher magnetic fields and improved instrumentation are opening up opportunities for membrane protein structural characterization both in solution and in solid-state NMR. The structural characterization of membrane proteins presents an additional challenge compared to water soluble proteins in that they have a unique orientation with respect to the their environment. How this environment and various models of this environment affect the membrane protein structure is still largely unknown. Most membrane proteins based on secondary structure predictions appear to be α-helical, although families of β-barrels, such as porins, are dominated by β-strands in the transmembrane domain. The α-helical proteins range from monotopic to polytopic with as many as 18 helices per protein monomer.

Integral membrane proteins are unique – they have their own structural families, their amino acid content is different from water soluble proteins and their external surfaces have a large hydrophobic region corresponding to the hydrophobic lipid domain [3]. The scarcity of water has numerous implications for this domain. Hydrophobic interactions are not considered to be important for the structural stability in these domains [4,5]. Water competes for the protein's intramolecular hydrogen bonds, thereby destabilizing them and inducing dynamics in the structural elements exposed to an aqueous environment [6]. Furthermore, such induced flexibility promotes curvature in an exposed helix. Of course, hydrophilic residues from asparagine to lysine as well as amino acids such as tryptophan and tyrosine rarely have their hydrophilic components buried in the hydrophobic environment. Because of this unique environment in which water is scarce and a low dielectric is present, the balance of interactions that stabilize protein structure is radically altered. In addition to the lack of hydrophobic interactions, electrostatic interactions are greatly enhanced to the extent that hydrogen bonds are difficult to break, both because of the low dielectic and because of the scarcity of catalytic water [6]. As a result current models for membrane protein folding have the secondary structural elements folding in the lipid-water interface followed by insertion across the membrane [7]. In other words, much of membrane folding occurs in an environment different from the native environment of the functional protein. This raises the likelihood that kinetically trapped states may be formed. Indeed, such states have been observed [8].

In this paper, to demonstrate both the characterization of topology and the high resolution obtainable, we have used the transmembrane peptide from M2 protein of Influenza A virus. This protein has a single transmembrane helix and forms a pH activated H^+ channel in the viral coat [9]. Even the M2 transmembrane peptide (M2-TMP, residues 22-46) forms H^+ channels and is blocked by the antiviral drug amantadine [10].

Solid-state NMR of uniformly aligned samples provides numerous opportunities to collect orientational constraints from various anisotropic spin interactions. These restraints have been shown to be very precise resulting from the high degree of alignment that can be achieved (a mosaic spread of as little as 0.3° has been characterized; [11]) leading to narrow linewidths relative to the breadth of the anisotropy. Here, just the anisotropic ^{15}N chemical shift and ^{15}N-^1H dipolar interactions will be observed in the polypeptide backbone through PISEMA (Polarization Inversion Spin Exchange at the Magic Angle) spectra [12]. This experiment results in high resolution spectra, not only in the chemical shift dimension, but also in the dipolar dimension.

Methods

M2-TMP was synthesized using solid-phase peptide synthesis and Fmoc chemistry on an Applied Biosystems 433A peptide synthesizer [13,14]. Oriented samples were prepared between glass plates (70 µm thick glass). Approximately 60 glass plates and 15 mg of peptide were used for the various samples leading to the data in Fig. 2&3 [15].

The solid-state NMR experiments were performed on a home-built 400 MHz spectrometer using a Chemagnetics data acquisition system and a wide-bore Oxford Instruments 400/89 magnet. A double resonance home-built probe with a rectangular coil was used to generate typical RF field strengths of 31 kHz for the cross-polarization match condition and 39 kHz for the Lee-Goldburg condition.

The PISEMA spectra of different structural models were simulated from their coordinates. Average values based on experimental data obtained from chemical shift powder patterns of single site labeled M2-TMP (σ_{11}=31.3, σ_{22}=55.2, σ_{33}=201.8 ppm relative to $^{15}NH_3NO_3$ at 0 ppm) were used as well as a motionally averaged value for the dipolar magnitude of 10.735 kHz. This latter value represents modest librational averaging of the static dipolar interaction based on an N-H bond length of 1.024 Å [16]. A typical relative orientation between the σ_{33} chemical shift tensor element and $\nu_{\parallel}$ of the dipolar tensor of 17° was used. PISA wheels were described as in Wang et al. [17] and Marassi & Opella [18]. The refinement was conducted using TORC (Total Refinement of Constraints; [19,20]) as recently applied [21,22].

Results and Discussion

Solid state NMR presents a unique opportunity to gain topological information on membrane proteins. The anisotropic spin interactions that form the axes of the

PISEMA spectra relate the orientation of spin interaction tensors to the magnetic field axis fixed in the laboratory frame of reference. Therefore, it is possible to define the magnitude of the laboratory Z axis component in the spectrum. For instance, a minimal Z component is presented in Fig. 1 orthogonal to the spectral plane of the anisotropic spin interactions near their maximal values and the Z component is maximal when the unique or nearly unique spin interaction tensor elements are perpendicular to B_0. Such dependence is shown for a line through the center of the simulated PISA wheels in the PISEMA spectrum.

This is not a novel result, but what is unique is the apparent representation of the X and Y axes in this spectrum. Such a result may seem to be counter intuitive because the NMR observables relate the tensors only to the Z axis and not to the X and Y axes of the laboratory frame. However, we know that the spin interactions are fixed in the molecular frame and that secondary structural elements position one peptide frame relative to another in Cartesian coordinates. Furthermore, since the ^{15}N-^{1}H dipolar and ^{15}N chemical shift tensors are not collinear these unique relative orientations of the peptide planes in three dimensional space yield unique spectral observables reflecting not only the Z axis components but also information on the laboratory frame X and Y axes. Consequently, X and Y axes can be defined in the spectral frame of reference for specific secondary structural elements. If the helix is tilted with respect to the Z axis in the ZY plane then the sets of axes in Figure 1 can be defined at a tilt of 38°. The X axis lies in the plane of the spectrum and the Y component in this plane is also displayed.

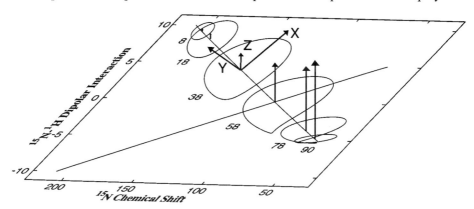

Fig. 1: Through the observation of PISA wheels in PISEMA spectra it is possible to define the laboratory frame of reference as a function of position in the spectrum. A series of Z axis components is displayed along the line formed by a string of PISA wheel centers reflecting the orientation of the dipolar and chemical shift unique (or nearly unique) tensor components relative to the magnetic field axis. The tilt of the α-helix for the PISA wheel calculation is defined to be in the ZY plane of the laboratory frame and consequently the in-plane Y axis components essentially parallel the centers line. The X axis is in the spectral plane nearly orthogonal to the centers line.

The orientation of the X axis and Y components varies with position in the spectrum as does the magnitude of the Y component. The result is a transformation of the laboratory frame of reference into the PISEMA spectral frame. This is a unique observation in NMR spectroscopy and consequently, projections of α-helices oriented with respect to the Z axis are imaged in the PISEMA spectrum.

To demonstrate such an image, data from the transmembrane helix of M2 protein is presented in Fig. 2A and compared with a helical wheel in Figure 2B. Here only one of the dipolar transitions is presented since the second one is redundant. Clearly, there is a close correlation between the image in the PISEMA spectrum, referred to as a PISA wheel (for Polar Index Slant Angle) and the helical wheel. The image provides both unique and important topological information. Not only does the image unequivocally determine that this segment of M2 protein is α-helical, but both the position, size and shape of the pattern documents that the helix is tilted at 38° with respect to the magnetic field and the normal to the lipid bilayer. In addition, the rotational orientation is quantitatively defined with respect to the top of the helix, defined by the peptide plane being normal to the YZ plane. Such data provides an excellent topological view for this polypeptide.

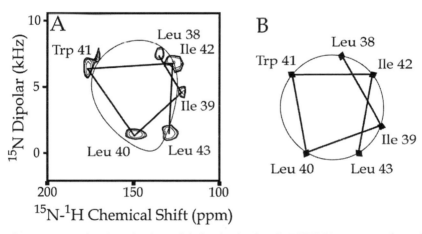

Fig. 2: Comparison of PISA wheels and helical wheels. A) PISEMA spectra of single and multiple site labeled samples for residues 38 - 43 from M2-TMP have been superimposed. The samples were uniformly aligned DMPC bilayers containing M2-TMP (16:1 molar ratio) oriented such that the bilayer normal was parallel to the magnetic field axis. Hydration was approximately 50%, pH was approximately 7 and samples were observed at room temperature above the gel to liquid crystalline phase transition temperature. The scaling of the dipolar interaction (0.81) by the PISEMA pulse sequence [12] was compensated for in drawing the dipolar axis. In other words, the dipolar scale has ben corrected by a factor of 0.81 as if the experiment did not scale the interaction. Only one of the symmetric pair of dipolar resonances is shown. B) The helical wheel (3.6 residues per turn) has been rotationally oriented to be in agreement with the PISA wheel.

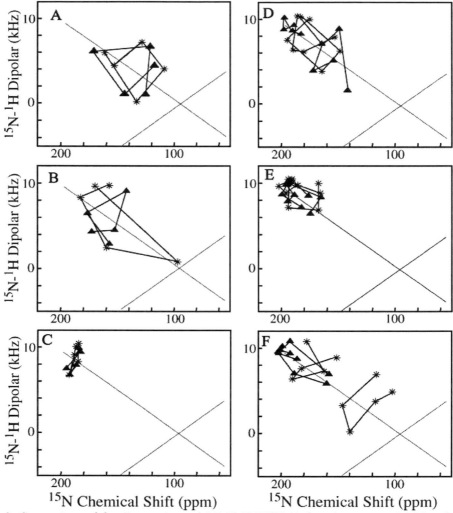

Fig.3: Comparison of the A) experimental M2-TMP PISEMA data (resonances 26-30 and 39-43) with simulated PISEMA spectra (residues 27-31 and 39-43) from various published models of the transmembrane helix of this M2 protein. B) From Kukol et al. [23], this structure is based on limited Infrared linear dichroism orientational constraints and a constant separation of helical axes resulting in a coiled coil structure. C) From Pinto et al. [24] where cysteine mutagenesis of the intact protein was used for restraints. D) Zhong et al. [25], snap-shot from a molecular dynamics simulation using an octane/water environment. E) Forrest et al. [26] snapshot from a molecular dynamics simulation using a water/POPC environment. F) Schweighofer & Pohorille [27] snap-shot from a molecular dynamics simulation using a water/DMPC environment. D-F) The molecular dynamics models are not symmetric structures and consequently PISEMA simulations have been performed on two of the four helices, but in fact, all four helices are different.

The presence of a clear PISA wheel pattern requires that the helical structure be uniform on the NMR microsecond time averaged scale. Indeed, the tilt of the peptide planes within the helix do not vary in their tilt relative to the helix axis by more than $\pm 2°$ indicating a very uniform structure. In addition, variations in the chemical shift tensor element magnitudes of more than $\pm 5\%$ are ruled out as are variations in the relative orientation of the principle elements of the tensors by more than $\pm 3°$.

Figure 3 presents a comparison of various models for this segment of the M2 protein, by simulating the PISEMA spectrum from coordinates of the various models. Fig 3A is the experimental data for residues 26 - 30 and 39 - 43. Arkin and coworkers restrained the helical tilt using orientational restraints from Infrared linear dichroism of ^{13}C carbonyl single site labeled M2 transmembrane peptide (Fig. 3B). The coiled coil structure is not consistent with the solid state NMR experimental data, but the tilt of the helix is very similar to that defined by the extensive solid state NMR data set. Pinto and coworkers developed a model based on cysteine mutagenesis and crosslinking in the intact protein (Fig. 3C). Such experimental data, while defining the rotational orientation well, did not accurately restrain the tilt of the helix.

Several molecular dynamics efforts have led to snap-shot structural models (Fig. 3D-F). The data obtained by NMR represents time averaged restraints where the averaging is on the time scale of the anisotropic spin interactions (microseconds). From a functional perspective this is important since ions, water molecules, ligands, etc., have time constants for their motion on the nanosecond to microsecond time scales. In other words, the molecules and ions that interact with the membrane proteins interact with a time averaged protein structure as opposed to a fleeting conformation on the femtosecond time frame. Local distortions in the snap-shot structures are clearly averaged on the microsecond time scale of the NMR experiment. In addition, the tetrameric bundle is clearly symmetric on the experimental time scale while the snap-shots are far from symmetric, not only in local distortions, but also in helical tilt. It is unlikely that such changes in tilt can be averaged on the microsecond time scale and consequently much of the scatter in the simulated PISEMA spectra (fig. 3D-F) is not likely even on the femtosecond time scale.

The uniformity of transmembrane protein helices is still an open question. Clearly, the increased influence of electrostatic interactions in the low dielectric environment will reduce the variability in α-helical hydrogen bond geometry seen on the surface of water soluble proteins. Unfortunately, there are few high resolution transmembrane protein structures. Shown in Fig. 4 are simulated PISEMA spectra for helix 1 (residues 12 through 30) and helix 4 (residues 104 through 125) of bacteriorhodopsin at different stages of refinement. The 3.5 and 2.35Å resolution structures show poorly defined PISA wheels in which we anticipate a roughly circular pattern with, on average, 3.6 resonances per cycle around the pattern (except near 90° tilt angles where there is just 1.8 resonances per cycle) [17,18]. These patterns are much better defined in the 1.7Å crystal structure which was published in June, 2000. The

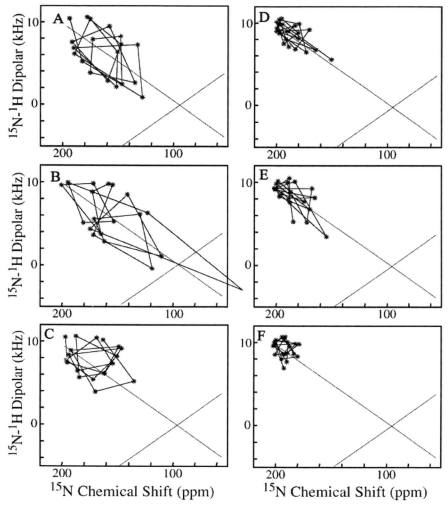

Fig. 4: PISEMA spectral simulations of A-C) helix 1 (residues 12-30) and D-F] helix 4 (residues 104-125) from X-ray structural characterizations of bacteriorhodopsin. A&D] 3.5Å resolution structure, PDB #1BM1 [28]. B&E) 2.35Å resolution structure, PDB #1AP9 [29]. C&F) 1.7Å resolution structure, PDB #1F50 [30].

improvement in the PISA wheels documents that the transmembrane helical structure in bacteriorhodopsin has nearly ideal geometry.

These results are confirmed in Figure 5 where the Rhamachandran diagrams as a function of structural resolution are shown. Again, the ideality of the helices is demonstrated as the distribution of the phi/psi torsion angles is reduced with improved resolution. All of this suggests that the refinement protocol and/or force field used in the

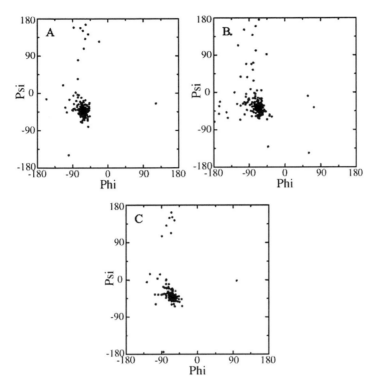

Fig. 5: Rhamachandran diagrams for all the helices in various bacteriorhodopsin models. A) 3.5Å resolution structure, PDB #1BM1 [28]. B) 2.35Å resolution structure, PDB #1AP9 [29]. C) 1.7Å resolution structure, PDB #1F50 [30].

refinement is not optimized for the altered balance of molecular interactions in the bilayer environment. With better definition of the PISA wheels, the α-helices of bacteriorhodopsin are not only better defined at higher resolution, but the helices become much more uniform with greatly reduced local distortions. Consequently, it can be anticipated that helices such as those in the KcsA K$^+$ channel at 3.2 Å resolution which predicts greatly distorted helices in the membrane environment will, in fact, be much less distorted when a high resolution structure is determined.

This is an important issue because it addresses the question of whether PISA wheel patterns will be recognizable in membrane proteins. Clearly, the pattern of resonances is well defined in M2-TMP (fig. 2 & 3) and has been easily recognized in uniformly ^{15}N labeled proteins such as colicin E1 [31], but spectral simulations from low resolution protein structures, such as KcsA typically predict spectra that are much more disordered than what has been observed (fig. 6). Based on the very positive simulation results from bacteriorhodopsin, we have reason to expect that such wheels will be recognizable in other membrane proteins and consequently topological

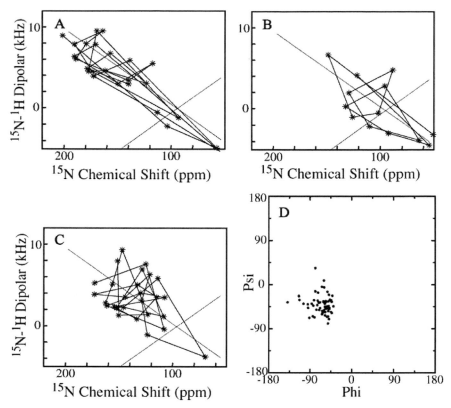

Fig. 6: PISEMA spectral simulations of A) helix 1 (residues 27-51), B) helix 2 (residues 62-74) and C) helix 3 (residues 86-112) from the X-ray structural characterization of the KcsA channel at 3.2 Å resolution, PDB #1BL8 [32]. D) The Rhamachandran diagram for these helices.

information will be available from solid-state NMR even without extensive resonance assignments.

While topological information is achievable, it is only the beginning of what can be gained from solid-state NMR-derived orientational restraints. Once complete assignments are made, refinement against these restraints, hydrogen bond distances, and a force field such as CHARMM can lead to a high resolution structure. The simulated annealing protocol developed for this purpose (TORC, [19,33]) must be able to search considerable conformational space, including global orientations, since the orientation of membrane proteins relative to their environment is critically important. Furthermore, search of global orientational as well as local conformational space is important because the solid-state NMR restraints orient the molecular frame relative to the laboratory Z axis. Such a restraint is an absolute structural restraint as opposed to the measurement of distances between sites within the protein leading to a relative restraint.

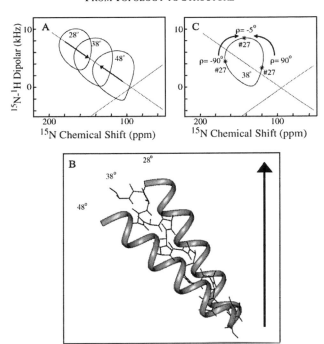

Fig. 7: Refinement without rigid body moves of the global orientation of M2-TMP. A) Illustrated in a simulated PISEMA spectrum, refinement from a helix tilt of 28° and 48° generated identical structures tilted at 38°. B) Ribbons represent two initial structures and the refined structures shown with backbone geometry superimpose. C) Illustrated in a simulated PISEMA spectrum, refinement from a helix rotational orientation of -90° and 90° generated identical structures rotated at -5°.

Shown in Fig. 7 are examples of the conformational search performed with the data from M2-TMP backbone. The search has been conducted without rigid body moves, but with compensating Ψ_i and ϕ_{i+1} moves of equal magnitude and opposite sign. In this way significant motion of the peptide plane can be induced without incurring prohibitive penalties from misalignment of the helix. It is demonstrated here that initial helix tilts of ± 10° from the observed value lead to identical refined structures. Similarly, it is possible to change the rotational orientation of the helix by ± 90° and achieve the same refined orientation for the helix. The initial structure developed from the PISA wheels has an error in tilt of less than ± 5° and in rotational orientation of less than ± 15°. Clearly, the conformational search more than covers the region of spatial ambiguity.

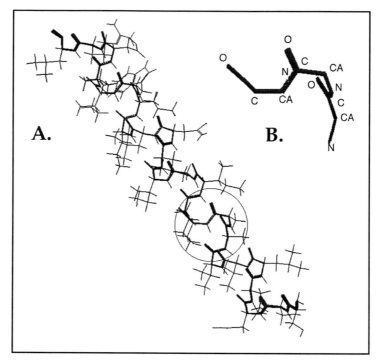

Fig. 8: A set of 30 backbone structural refinements of M2-TMP illustrating the very high resolution of the structure (0.05Å heavy atom rmsd for the backbone). An expansion of residues 29-31 in the backbone shows the remarkable quality of the refinement. Also displayed is one set of side chain conformations that have been energy minimized.

High resolution refinement is achieved through introducing a small random diffusion constant for the atom moves. In this way the backbone geometry is not limited to ideal values. Balancing the experimental restraints relative to the CHARMM energy is critically important. Recently, the gramicidin A structure, the first membrane bound structure to be determined by solid-state NMR was validated [34] and complete cross validation has also been demonstrated [22]. This latter procedure resulted in a unique weighting factor for the CHARMM vs. experimental restraints such that overfitting and underfitting the data could be avoided. This protocol has led to a very high resolution structure for the transmembrane helix of M2-TMP (Fig 8). The backbone heavy atom rms deviation for thirty refinements is 0.05Å.

In solution NMR it is customary to have an average of 15 or more restraints per amino acid residue to adequately restrain the macromolecular structure. Many of these restraints involve the side chain, but while few NOEs are obtained for the backbone, other restraints such as chemical shifts, torsional angles and residual dipolar interactions lead to a large number of backbone restraints. Here just two precise restraints are required in addition to the PISA wheel characterization of the α-helix. Moreover, the

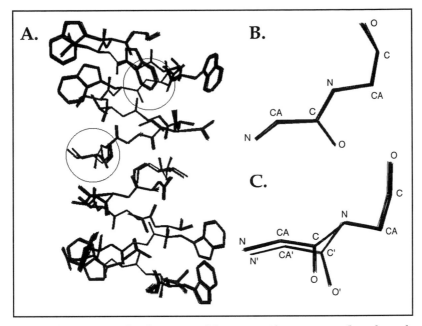

Fig. 9: A set of 40 structural refinements of the gramicidin monomer (but shown here as the native dimer) again illustrating the very high resolution of another solid-state NMR derived structure. Through the high resolution refinement, it was possible to detect two conformational substates as shown in an expansion of the amino terminus (formyl-Val₁ backbone). These substates are consistent with the observed single-valued orientational restraints for this part of the structure and the two well resolved isotropic chemical shifts observed for this Val₁ carbonyl carbon. Such conformational ambiguities do not occur elsewhere in the structure.

refinement against the solid-state NMR restraints leads to a very high resolution structure. While many additional restraints are possible, such as ^{15}N-^{13}C dipolar interactions, ^{13}C and ^{1}H chemical shifts, these restraints are not necessary. Side chain conformation can also be achieved as demonstrated with site specific labeling in gramicidin A [33].

It is important to recognize that this solid-state NMR-derived structure is a time-averaged structure. The high resolution structural characterization of the backbone does not imply that the structure is rigid, indeed significant backbone motion leads to relatively short ^{15}N relaxation times. While the high resolution structure conceals the dynamic fluctuations it does not mask structural ambiguities or conformational substates. In the high resolution gramicidin A structure the formylated amino termini at the bilayer center appears to have two conformational substates (fig. 9). Magic angle spinning observation of the ^{13}C carbonyl signal of Val₁ shows two values for the isotropic chemical shifts and consequently, the interconversion between these states is on the 10 ms timescale or slower. Such conformational substates may be related to channel flicker or open channel noise. It is such high resolution detail that will lead to a solution for some of the most important questions in electrophysiology.

Conclusions

Solid-state NMR is maturing into a technology that can characterize membrane protein structure from topology to the highest resolution yet achieved by any approach for membrane proteins. A complete high resolution structure of a pentadecapeptide, gramicidin A, has been characterized in the liquid crystalline phase of a lipid bilayer. The backbone structure of the transmembrane helix from M2 protein has been characterized in a similar environment. In looking forward to the structural characterization of intact membrane proteins we can anticipate that many helices will be nearly uniform in structure giving rise to observable PISA wheels in solid state NMR spectra.

Acknowledgments

The authors are indebted to Professors Klein, DeGrado, Sansom, Arkin and Pohorille as well as their colleagues for sending us coordinates of their models of M2-TMP. This work was supported by the National Science Foundation, DMB 9986036 and the work was largely performed at the National High Magnetic Field Laboratory supported by Cooperative Agreement, DMR 9527035 and the State of Florida.

References:

[1]   Fu, R. and Cross, T.A., Ann. Rev. Biophys. Biomol. Struct. 28 (1999) 235.
[2]   North, C.L. and Cross, T.A., Biochemistry 34 (1995) 5883.
[3]   von Heijne, G. EMBO J. 5 (1986) 3021.
[4]   Popot, J.L. and Engelman, D.M., Biochemistry 29 (1990) 4031.
[5]   Engelman, D.M. and Steitz, T.A. in *The Protein Folding Problem* (ed. Wetlaufer, D.B.) Westview, Boulder, Colorado, (1984) 87.
[6]   Xu, F. and Cross, T.A., Proc. Natl. Acad. Sci. U.S.A. 96 (1999) 9057.
[7]   Engelman, D.M., Science 274 (1996) 1850.
[8]   Arumugam, S., Pascal, S., North, C.L., Hu, W., Lee, K.-C., Cotten, M., Ketchem, R.R., Xu, F., Brenneman, M., Kovacs, F., Tian, F., Wang, A., Huo, S. and Cross, T.A., Proc. Natl. Acad. Sci. U.S.A. 93 (1996) 5872.
[9]   Lamb, R.A., Holsinger, L.J. and Pinto, L.H., In Wemmer, D.E. (ed.) Receptor-Mediated Virus Entry into Cell, Cold Spring Harbor Press: Cold Spring Harbor, New York (1994) p. 303.
[10]   Duff, K.C. and Ashley, R.H., Virology 190 (1992) 485.
[11]   Cross, T.A., Ketchem, R.R., Hu, W., Lee, K.-C., Lazo, N.D. and North, C.L. Bull. Magn. Reson. 14 (1992) 96.
[12]   Wu, C.H., Ramamoorthy, A. and Opella, S.J., J. Magn. Reson. 109A (1994) 270.
[13]   Fields, C.G., Fields, G.B., Nobles, R., Cross, T.A., Int. J. Peptide Protein Res. 33 1989) 298.
[14]   Kovacs, F. and Cross, T.A., Biophys. J. 74 (1977) 2511.

[15] Song, Z., Kovacs, F.A., Wang, J., Denny, J.K., Shekar, S.C., Quine, J.R. and Cross, T.A., Biophys. J. 79 (2000) 767.
[16] Wang, J., Denny, J., Tian, C., Kim, S., Mo, Y., Kovacs, F., Song, Z., Nishimura, K., Gan, Z., Fu, R., Quine, J.R. & Cross, T.A., J. Magn. Reson. 144 (2000) 162.
[17] Cross, T.A. and Quine, J., Concepts in Magn. Reson. 12 (2000) 55.
[18] Marassi, F.M. and Opella, S.J., J. Magn. Reson. 144 (2000) 150.
[19] Ketchem, R.R., Hu, W. and Cross, T.A., Science 261 (1993) 1457.
[20] Ketchem, R.R., Roux, B. and Cross, T.A., In Merz, K.M. and Roux, B. (eds.) Membrane Structure and Dynamics, Birkhauser, Boston, MA, (1996) p. 299.
[21] Kovacs, F.A., Denny, J.K., Song, Z., Quine, J.R. and Cross, T.A., J. Mol. Biol. 295 (2000) 117.
[22] Kim, S., Quine, J.R. and Cross, T.A., J. Am. Chem. Soc., in press.
[23] Kukol, A., Adams, P.D., Rice, L.M., Brunger, A.T. and Arkin, I.T., J. Mol. Biol. 286 (1999) 951.
[24] Pinto, L.H., Dieckmann, G.R., Gandhi, C.S., Papworth, C.G., Braman, J., Shaughnessy, M.A., Lear, J.D., Lamb, R.A. and DeGrado, W.F., Proc. Natl. Acad. Sci. U.S.A. 94 (1997) 11301.
[25] Zhong, Q., Husslein, T., Moore, P.B., Newns, D.M., Pattnaik, P. and Klein, M.L., FEBS Letters 434 (1998) 265.
[26] Forrest, L.R., Kukol, A., Arkin, I.T., Tieleman, D.P. and Sansom, M.S.P., Biophys. J. 78 (1999) 55.
[27] Schweighofer, K.J. and Pohorille, A., Biophys. J. 78 (2000) 150.
[28] Sato, H., Takeda, K., Tani, K., Hino, T., Okada, T., Nakasako, M., Kamiya, N. and Kouyama, T., Acta Crystallogr. Biol. Crystallogr. D55 (1999) 1251.
[29] Pebay-Peyroula, E., Rummel, G., Rosenbusch, J.P. and Landau, E.M., Science 277 (1997) 1676.
[30] Luecke, H., Schobert, B., Cartailler, J.P., Richter, H.T., Rosengarth, A., Needleman, R., Lanyi, J.K., J. Mol. Biol. 3000 (2000) 1237.
[31] Kim, Y., Valent, K., Opella, S.J., Schendel, S.L. and Cramer, W.A. Prot. Sci. 7 (1998) 342.
[32] Doyle, D.A., Morais-Cabral, M., Pfuetzner, R.A., Kuo, A., Gulbis, J.M., Cohen, S.L., Chait, B.T., Science 280 (1998) 69.
[33] Ketchem, R.R., Roux, B. and Cross, T.A., Structure 5 (1997) 1655.
[34] Kovacs, F., Quine, J. and Cross, T.A., Proc. Natl. Acad. Sci. USA 96 (1999) 7910.

Toward dipolar recoupling in macroscopically ordered samples of membrane proteins rotating at the magic angle

**Clemens Glaubitz*, Marina Carravetta, Mattias Edén,
Malcolm H. Levitt**

Physical Chemistry Division, Arrhenius Laboratory
Stockholm University, S-10691 Stockholm, Sweden
*e-mail: glaubitz@physc.su.se

Abstract

MAS NMR spectroscopy can be combined with the advantages of uniaxially ordered samples of membrane proteins as demonstrated in the so-called MAOSS (magic angle oriented sample spinning) experiment. Under these conditions, dipolar recoupling methods can be used to determine the orientation of internuclear vectors with respect to the MAS rotor frame. However, most approaches to measure dipolar couplings yield angle ambiguities even in the static, non-spinning case. Here, we present the possibility of overcoming these problems by deriving a new homonuclear double-quantum radio frequency pulse sequence based on an eightfold symmetry. Only dipolar Hamiltonian terms with spatial components m=±2 are recoupled with high efficiency allowing unambiguous angle determinations. Preliminary data demonstrate the applicability of this experiment to oriented samples.

Introduction

Magic angle spinning (MAS) is an essential NMR technique for studying disordered samples of biological materials such as biomembranes [1]. The large variety of applications involves, for example, investigations of non-covalent ligands bound to transport proteins and receptors [2,3], the possibility of studying the location of protein segments within the membrane [4] and structural studies of isotope labelled, immobilized proteins [5,6].

The structural description of membrane proteins is technically challenging since data about secondary and tertiary conformation as well as about location and orientation

S.R.Kiihne and H.J.M.deGroot (eds.), Perspectives on Solid State NMR in Biology, 71-81.
©2001 Kluwer Academic Publishers. Printed in the Netherlands.

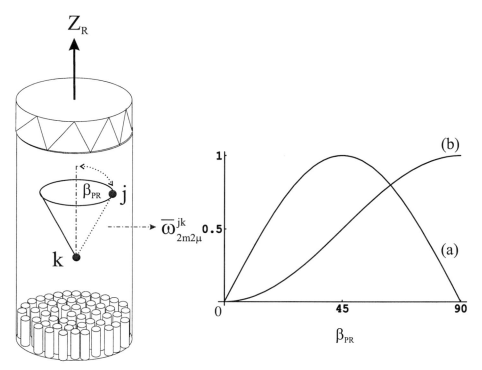

Fig.1: Dipolar recoupling methods can be used to determine the orientation β_{PR} of the internuclear vectors between spins j and k with respect to the MAS rotor axis Z_R in an ordered sample (such as a macroscopically oriented membrane protein). However, most homo- as well as heteronuclear recoupling techniques used so far recouple terms of the dipolar Hamiltonian with spatial elements $m=\pm 1$ which are scaled with $\sin 2\beta_{PR}$ leading to angle ambiguities (a). Here it is shown that specific recoupling sequences can be designed which only use dipolar Hamiltonian terms with $m=\pm 2$ ($\sin^2 \beta_{PR}$) allowing an unambiguous determination of the orientation of the dipolar coupling tensor (b)

in the anisotropic membrane environment have to be obtained. The latter question is best approached by using macroscopically ordered samples. This is routinely done for membrane-bound peptides using static NMR techniques [7,8]. Macroscopically ordered samples with well oriented lipids and proteins can be prepared either by aligning them as bilayers on glass disks [9,10] or by using bicelles [11-13]. Since the signal frequency observed from second rank tensor interactions such anisotropic chemical shift, and quadrupolar and dipolar couplings depends on its orientation with respect to the field of interaction, sharp resonance lines would be observed if all tensors had the same orientation with respect to the magnetic field B_0. However, the distribution of local geometry axes relative to the average alignment axis (mosaic spread), which is common for large proteins such as rhodopsin or purple membrane, causes a significant contribution to the linewidth, making this static NMR approach less straightforward [14].

A solution to that problem has been demonstrated in the form of a new hybrid technique by applying MAS NMR spectroscopy to oriented membrane protein samples (so called MAOSS - magic angle oriented sample spinning - experiment) [15]. The general possibility of obtaining orientation distribution functions by MAS NMR has been shown before for ordered polymers and DNA fibres [16,17]. The principal experimental idea is shown in Fig.1. A uniformly aligned lipid/protein film fixed on thin glass disks is mounted into a MAS rotor, which is subject to rotations at modest speeds about the magic angle in the magnetic field [18]. The principal axis system of the chosen NMR interaction (dipolar or quadrupolar coupling or chemical shift anisotropy) is related to the protein structure by a set of Euler angles, which depends on a number of parameters (local conformation, hydrogen bonds, etc.). The molecular reference system itself is related to the sample director frame of the glass disks by another set of Euler angles which account for the two dimensional distribution and disorder effects [18,19]. The director frame and the MAS rotor fixed reference frame are identical. The MAOSS experiment, which can be seen as a complementary technique to MAS studies on isotropically ordered systems, has already been applied to a number cases. Examples include the determination of the ligand orientations in bacteriorhodopsin [19] and bovine rhodopsin [20] as well as applications to peptides and lipids [15,21]. First experiments on oriented, multiply labelled samples of ^{15}N-Met-bacteriorhodopsin showed that in principle many angular constraints can be obtained simultaneously by deconvoluting the ^{15}N MAS sideband pattern (Glaubitz, C., Mason, J., Watts, A., unpublished results).

In all these cases – as in static NMR experiments using oriented samples - the anisotropies of deuterium quadrupolar or ^{13}C and ^{15}N chemical shift interactions were utilized. Analysing these data in terms of molecular structure however requires some knowledge about the tensor orientations within the molecular reference frame. Usually, the secondary structure has to be known, especially in order to interpret the orientation of ^{15}N CSA tensors [22,23].

A more direct approach would be possible by combining MAS on oriented samples with dipolar recoupling techniques. Selectively reintroducing the dipolar coupling while maintaining the high spectral resolution achieved by MAS would allow direct determination of the tilt angle β_{PR} of the internuclear vector between spins j and k with respect to the sample/rotor reference system (Fig.1). However, most homo- and heteronuclear recoupling techniques used today feature a more or less complex orientational dependence on the Euler angles Ω_{PR}. For example, the sequence C7 [24] depends on $\sin 2\beta_{PR}$ while REDOR [25] and MELODRAMA [26] additionally depend on $\cos\gamma_{PR}$. The only exception described so far is rotational resonance at the n=2 condition [27]. However, this method, generally requiring high spinning frequencies, is limited to samples with appropriate differences in isotropic chemical shifts and is sensitive to interference from the chemical shift anisotropies.

Here, we demonstrate the design of specific recoupling schemes that only recouple terms with $\sin^2\beta_{PR}$ dependence, allowing unambiguous angle determination. The method is broadband with respect to isotropic shifts, is robust with respect to chemical shift anisotropy and does not require very rapid sample spinning.

Theory

It has been shown in the past that symmetry arguments can be exploited to design rf pulse sequences for tasks like homonuclear recoupling or heteronuclear decoupling in the presence of sample rotation [24,28-30]. Two symmetry classes, denoted as CN_n^v and RN_n^v, were discovered [28,30]. Sequences based on the symmetry class CN_n^v are specified by three integer numbers N,n,v with the following properties: (a) they consist of phase-shifted repetitions of a rf cycle C, each of which returns the irradiated spins to their initial state (in the absence of other interactions), (b) each cycle has the duration $\tau_C=(n\,\tau_r/N)$ with $\tau_r=|2\pi/\omega_r|$, i.e. N rf cycles span n rotor periods, and (c) the phase of the qth cycle C_q is given by $\Phi_q=2\pi vq/N$ (q=0,1,2...N-1). The following selection rules for the average Hamiltonian components were derived in reference [28]:

$$\overline{H}_{lm\lambda\mu}^{(1)} = 0 \text{ if } mn - \mu v \neq N \times \text{integer number} \tag{1}$$

$$\overline{H}_{l_1m_1\lambda_1\mu_1:l_2m_2\lambda_2\mu_2}^{(2)} = 0 \text{ if } \begin{cases} m_1n - \mu_1v \neq N \times \text{integer number} \\ m_2n - \mu_2v \neq N \times \text{integer number} \\ (m_2 + m_1)n - (\mu_2 + \mu_1)v \neq N \times \text{integer number} \end{cases} \tag{2}$$

A particular spin interaction is classified by index l referring to its spatial rotational rank (with m=-l,-l+1,...,0,...,+l) and λ denoting the rotational rank of spin polarization rotations by the resonant rf field (with μ=-λ,-λ+1,...,0,...,+λ). In this notation, isotropic and anisotropic chemical shifts have l=0, λ=1 and l=2, λ=1, respectively. Homonuclear dipolar couplings are described by l=λ=2. These selection rules have been used successfully to construct a number of pulse sequences e.g. to perform double-quantum filtered dipolar recoupling using the symmetry $C7_2^1$ which has found widespread application. The simplified average Hamiltonian for $C7_2^1$ is given in [24] as

$$\overline{H}^{(1)} = \sum_{lm\lambda\mu} \overline{\omega}_{lm\lambda\mu}^{jk} T_{\lambda\mu}^{jk} = \overline{\omega}_{2122}^{jk} T_{22}^{jk} + \overline{\omega}_{2-12-2}^{jk} T_{2-2}^{jk} \tag{3}$$

where $T_{2\pm2}^{jk}$ are second rank spin operators for the interactions between spins j and k. Explicit expressions for the magnitude $\overline{\omega}_{lm\lambda\mu}^{jk}$ are given in references [24] and [28]. These terms are proportional to the reduced Wigner function $d_{0m}^l(\beta_{PR})$ which becomes

$$d_{0\pm1}^2 = \pm\sqrt{3/2}\sin 2\beta_{PR} \tag{4}$$

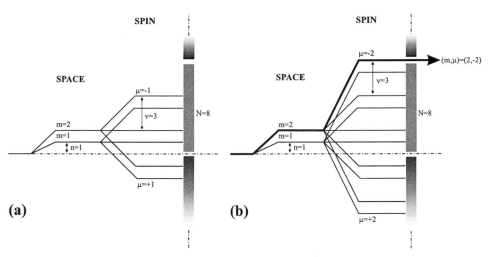

Fig. 2: Space-spin selection diagram (SSS diagram) for $C8_1^3$. All CSA modulation components are suppressed (a). One single 2Q dipole-dipole component with quantum numbers $(m,\mu)=(2,-2)$ is selected (b). The mirror images for $m-1$ and $m=-2$ have been suppressed for simplicity.

in equation (3) leading to angle ambiguities in ordered systems (Fig.1). This problem could be solved by analysing expressions (1) and (2) for symmetries which result in similar properties as $C7_2^1$ but with $m=\pm2$ to yield an average Hamiltonian of the form

$$\overline{H}^{(1)} = \overline{\omega}_{2222}^{jk}T_{22}^{jk} + \overline{\omega}_{2-22-2}^{jk}T_{2-2}^{jk} \tag{5}$$

Using $m=\pm2$, the magnitudes $\overline{\omega}_{lm\lambda\mu}^{jk}$ are scaled with

$$d_{0\pm2}^2 = \sqrt{3/8}\sin^2\beta_{PR} \tag{6}$$

leading to a monotonic angle dependence for $0\leq\beta_{PR}\leq90°$ (Fig.1b). At the same time, for optimal results, one has to ensure that chemical shift terms $\omega_{2m\lambda\mu}$ with $m\neq0$ are excluded, double quantum dipolar terms with $\mu=\pm2$ are selected, and all other dipolar terms are suppressed. Each double quantum term should be associated with only a single spatial component $m=2$ or $m=-2$. This eliminates the orientation dependence on γ_{PR} in the first order average Hamiltonian in favour of a maximum magnitude of the effective dipolar Hamiltonian while creating a monotonic orientation dependence on β_{PR} (see Fig.1).

   A set of 23 possible symmetries in the range of $1\leq N\leq20$, $1\leq n\leq5$, $0\leq v\leq10$ are found analysing theorems (1) and (2). Here, we restrict ourselves to one particular solution using an eightfold symmetry with $N=8$, $n=1$, $v=3$; denoted in the following as $C8_1^3$. The selection of particular space-spin components can be illustrated schematically using a space-spin selection diagram [28] as shown in Fig.2. Each level indicates a value of mn-µv, while the barrier symbolises the inequality of theorem (1) [28]. As shown in Fig.2a for CSA components (m={±1, ±2}, µ={0, ±1}) no pathway can be found to pass through the "selection wall". This means that all CSA components are suppressed by $C8_1^3$ at least in the first order average Hamiltonian. The fact that only homonuclear dipolar components with (m,µ)=(2,–2) and (–2,2) are allowed is shown in Fig.2b. The terms with µ=±2 indicate double quantum coherence. Furthermore, each term is only associated with m=±2 rotational components which creates the desired monotonic orientation dependence required for studies on ordered systems.

Methods

The symmetry principles visualized in Fig.2 are independent of the experimental implementation of $C8_1^3$. The optimal choice of the details for the C element depends on a number of factors such as robustness of the sequence with respect to rf field inhomogeneities or interference from isotropic chemical shifts. Additional constraints with respect to spinning speed and rf amplitudes are important in the case of MAOSS type of experiments. Usually, larger MAS rotors are necessary (6-7.5mm) for accommodating glass disks containing the protein sample. This limits power levels (up to 100kHz) and spinning rates (2-6kHz) to moderate levels, also due to sample stability aspects [18].

For the demonstration here, we have chosen $C_0=(2\pi)_0$ which is stepwise phase incremented by $3/4\,\pi$ as shown in Fig.3. Eight C elements span one rotor period. The shaded sequence elements have variable phases according to standard procedures for selecting signals passing through double quantum coherence [32]. The signal build-up can be measured for a variable double-quantum excitation time τ_E while keep the reconversion time τ_R constant [29]. Oscillations in the observed build-up curve reveal the dipolar coupling between the ^{13}C labeled spins. This approach has been shown to be especially useful in difficult situations, where the isotropic chemical shifts of the two spins are similar, such as in [10,11-$^{13}C_2$]-E-retinal [31].

The experimental situation encountered by performing MAS on oriented membrane proteins in MAOSS type of experiments (see Fig.1) can be conveniently emulated by placing an isotope labelled single crystal in the MAS rotor. The carousel symmetry, i.e. the two-dimensional distribution of the protein about the MAS rotor axis can be mimicked by not synchronising the data acquisition with the sample rotation. In this case, a slow-speed ^{13}C spectrum will only feature nearly absorptive sidebands in contrast to rotor-synchronized sampling [33,34]. A single crystal of [1,2-$^{13}C_2$]-glycine (4x4x3mm^3), covered in teflon tape, was placed in the centre of a 4mm Chemagnetics MAS rotor. The crystal position was adjusted to allow stable sample rotation. To compare the DQF performance of $C8_1^3$ with respect to isotropic chemical shifts, a polycrystalline sample of [10,11-$^{13}C_2$]-E-retinal was also used.

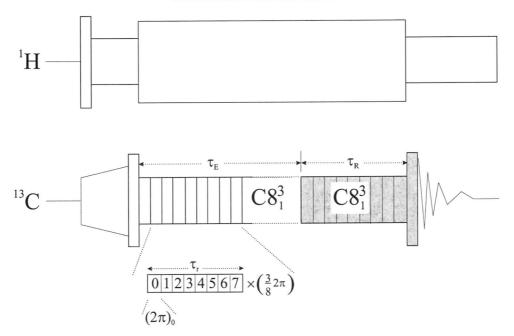

Fig. 3: Radio frequency pulse sequence for double quantum filtered measurements of dipolar couplings using $C8_1^3$. Eight rf cycles are timed to occupy one rotational period. Each cycle consists of one 2π pulse with a $3/4 \, \pi$ phase increment. It has been shown, that distance data can be obtained by measuring the DQF efficiency as a function of the DQ excitation period τ_E while keeping the reconversion time τ_R constant [31].

All experiments were performed using a Chemagnetics 200 Infinity spectrometer and a 4mm triple resonance probe. Experimental details are given in the legend to Fig. 4.

Results and Discussion

The double-quantum filtering performance of $C8_1^3$ is shown in Fig.4 for $[1,2-^{13}C_2]$-glycine (a) and $[10,11-^{13}C_2]$-E-retinal (b). Conventional cross-polarized spectra at a sample rotation rate of $\omega_r/2\pi$=5kHz show both labelled sites in (a) and (b). The sample of $[10,11-^{13}C_2]$-E-Ret was diluted to 10% in unlabelled material, explaining the additional intensities. The top spectra are the result of passing the cross-polarized signal through double-quantum coherence using the sequence in Fig.3. Excitation and reconversion sequences were of the same length using q=24 elements. In both cases, ca 40% of the spin magnetization passed through the double-quantum filter. This shows, that $C8_1^3$ seems to be fairly robust with respect to differences in isotropic chemical shifts.

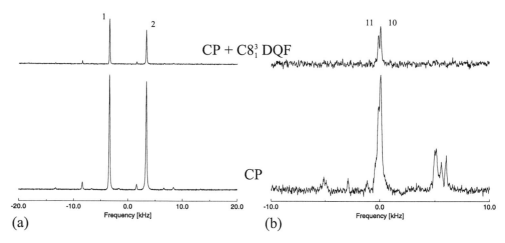

Fig. 4: MAS ^{13}C spectra of polycrystalline samples of [1,2-$^{13}C_2$]-glycine (a) and [10,11-$^{13}C_2$]-E-retinal (b) with and without double quantum filtering using $C8_1^3$. Spectra were acquired at $\omega_r/2\pi=5kHz$ with $\tau_E=600\mu s$, a CP contact time of 2ms and at 4.9 T. A proton decoupling power of 70 kHz during acquisition and 100 kHz while applying the $C8_1^3$ pulse-train was used. The (2π)-pulses had a length of 25 μs. Ca. 40% DQ efficiency is achieved in both cases.

One possibility for determining dipolar couplings using these symmetries is to measure the observed double-quantum efficiency as a function of the excitation time τ_E while keeping the reconversion sequences at constant length (Fig.3) [31]. The oscillations in the observed build-up curves are a sensitive measure for the strength of the dipolar coupling. Fig.5a shows the signal build-up for a polycrystalline sample and for a single crystal of [1,2-$^{13}C_2$]-glycine. Both data sets were obtained at $\omega_r/2\pi=5kHz$. The dipolar coupling obtained for the unoriented sample of 2.25kHz corresponds to a distance of 0.15nm which agrees well, within the error limits, with the literature value of 0.1543nm [36]. Possible error sources include intermolecular effects, since the sample was not diluted with unlabelled glycine, and vibration effects. It has also been shown, that MAS fails to completely eliminate the coupling between ^{13}C and ^{14}N, which has a spin I=1 and a 99.63% natural abundance [37]. The residual dipolar coupling is inverse proportional to the Larmor frequency and is an additional source of experimental error at lower magnetic fields [38]. The same experiment performed on the single crystal shows a totally different signal build-up with a much longer periodicity indicating a smaller dipolar coupling. The tilt angle between the C_α and the carboxyl nuclei and the MAS rotor axis β_{PR} was so determined to be 30°. The unit cell of glycine as crystallized here in the monoclinic space group P2$_1$/n, contains two pairs of molecules [35,36]. The molecules in each pair have a slightly different orientation, so that only a superposition of curves is observed. Also, the exact orientation of the crystal in the MAS rotor is not known, since its placement in the rotor was determined by its mechanical dimensions to ensure stable spinning rather than by its crystallographic

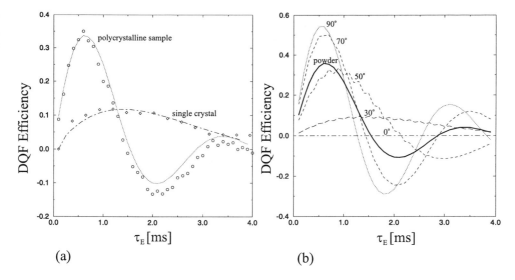

Fig. 5: Comparison of DQF build-up curves from a polycrystalline and a single crystal sample of 1-2-^{13}C-glycine using the sequence shown in Fig.3 (a). Data were acquired at $\omega_r/2\pi=5kHz$ with a fixed reconversion period ($\tau_R= 600$ and 1500 μs for polycrystalline and single crystal samples, respectively). The best fit for the single crystal build-up is obtained for $\beta_{PR}=30^o$. Simulated build-up curves demonstrate the monotonic dependence of the recoupled dipolar coupling from the tilt angle β_{PR} which allows unambiguous tilt angle measurements.

axes. However, it was possible to verify the obtained results by observing dipolar splittings while spinning the crystal at the rotational resonance conditions n=1 and n=2 at $\omega_r/2\pi$=6.94 and 3.47kHz (data not shown). Interestingly, at n=1 the average dipolar Hamiltonian depends on spatial components with m=±1 while at n=2 only m=±2 terms contribute to the observed dipolar coupling [27]. Therefore, an unambiguous additional angle determination was possible which led to the same result. The fact that only one dipolar splitting was observed at rotational resonance also indicates that both inequivalent molecules in the unit cell differ only slightly in their orientation, which is in agreement with the crystal structure [36].

Various simulated build-up curves over the range $0 \leq \beta_{PR} \leq 90°$ are shown in Fig.5b to illustrate the angle dependence of the dipolar coupling using $C8_1^3$.

Conclusions

We have shown that the orientation of homonuclear dipolar couplings in ordered samples can be determined unambiguously. This is achieved by using an eightfold symmetry $C8_1^3$, which selects only terms of the dipolar Hamiltonian scaled with $\sin^2\beta_{PR}$.

The orientation of dipolar couplings is directly related to the orientation of internuclear vectors. Hence, angular constraints can so be obtained which are directly related to the molecular structure. This is an advantage over studying CSA tensor orientations where additional information about their position in the molecular frame would be required in order to obtain the orientation of molecular segments.

It should also be pointed out that the unambiguous measurement of the dipolar coupling orientation is not possible in a static NMR experiment, unless data are acquired with different sample orientations in the static field as shown in early single crystal experiments [35]. This underlines the strength of the approach presented here with respect to the study of complex systems such as membrane proteins, where sensitivity and spectral resolution are an important concern.

Further detailed studies and applications of these principles to a uniformly aligned, isotope labelled membrane protein are currently in progress and will be reported elsewhere.

Acknowledgements

O. Johannesson, H. Luthmann, A. Brinkman, X. Zhao, P.K. Madhu, J. Schmedt auf der Günne, C. Hughes are acknowledged for technical help and discussions and Oleg Antzutkin for advice and preparing the single crystal sample.

References

[1]   Smith, S.O., Aschheim, K., Groesbeck, M., Q.Rev.Biophys. 29 (1996) 395.
[2]   Watts, A., Burnett, I.J., Glaubitz, C., Gröbner, G., Middleton, D., Spooner, P.J.R., Williamson, P.T.F., Eur.Biophys.J. 28 (1998) 84.
[3]   Glaubitz, C., Gröger, A., Gottschalk, K., Spooner, P., Watts, A., Schuldiner, S., Kessler, H., FEBS Lett. 480 (2000) 127.
[4]   Gröbner, G., Glaubitz, C., Watts, A., J. Magn. Reson. 141 (1999) 335.
[5]   van Rossum, B.-J., Steensgaard, D.B., Mulder, F.M., Boender, G.J., Schaffner, K., Holzwarth, A.R., de Groot, H.J.M., Biochemistry in press (2001)
[6]   Feng, X., Verdegem, P.J.E., Lee, Y.K., Sandstrom, D., Eden, M., BoveeGeurts, P., de Grip, W.J., Lugtenburg, J., deGroot, H.J.M., Levitt, M.H., J.Am.Chem.Soc. 119 (1997) 6853.
[7]   Opella, S.J., Nat.Struct.Biol. 4 (1997) 854.
[8]   Fu, R.Q., Cross, T.A., Ann.Rev.Biophys.Biomol.Struct. 28 (1999) 235.
[9]   Cross, T.A., Opella, S.J., Curr.Opin. Struct. Biol. 4 (1994) 574.
[10]   Gröbner, G., Taylor A., Williamson, P.T.F., Choi, G., Glaubitz, C., Watts, J.A., de Grip, W.J., Watts, A., Anal. Biochem. 254 (1997) 132.
[11]   Sanders, C.R., Hare B.J., Howerd, K., Prestegard, J.H., Progr.Nucl.Magn.Reson. 26 (1993) 421.
[12]   Prosser, S.R., Hunt, S.A., Dinitale, J.A., Vold, R.R., J. Am. Chem. Soc. 118 (1996) 269.
[13]   Howard, K.P., Opella, S.J., J. Magn. Reson. 112 (1996) 91.
[14]   Gröbner, G., Choi, G., Burnett, I., Glaubitz, C., Verdegem, P.J.E., Lugtenburg, J., Watts, A., FEBS Lett. 422 (1998) 201.

[15]   Glaubitz, C., Watts, A., J. Magn. Reson. 130 (1998) 305.
[16]   Harbison, G.S., Vogt, V.D., Spiess, H.W., J.Chem.Phys. 86 (1987) 1206.
[17]   Song, Z.Y., Antzutkin, O.N., Lee, Y.K., Shekar, S.C., .Rupprecht, A., Levitt,M.H., Biophys.J. 73 (1997) 1539.
[18]   Glaubitz, C., Conc.Magn.Reson. 12 (2000) 137.
[19]   Glaubitz, C., Burnett, I.J., Gröbner, G., Mason, A.J., Watts, A., J.Am.Chem.Soc. 121 (1999) 5787.
[20]   Gröbner, G., Burnett, I., Glaubitz, C., Choi, G., Mason, J., Watts, A., Nature 405 (2000) 810-813.
[21]   Glaubitz, C., Gröbner, G., Watts, A., Biochim. Biophys. Acta 1463 (2000).
[22]   Hartzell, C.J., Whitfield, M., Oas, T.G., Drobny, G.P., J.Am.Chem.Soc. 109 (1987) 5966.
[23]   Walling, A.E., Pargas, R.E., deDios, A.C., J.Phys.Chem. 101 (1997) 7299.
[24]   Lee, K.Y., Kurur, N.D., Helmle, M., Johannessen, O.G., Nielsen, N.C. and Levitt, M.H., Chem.Phys.Lett. 242 (1995) 242.
[25]   Gullion, T. and Schaefer, J., Adv. Magn. Reson. 13 (1989) 57.
[26]   Sun, B.Q., Rienstra, C.M., Costa, P.R., Williamson, J.R., Griffin, R.G., J.Am.Chem.Soc. 119 (1997) 8540.
[27]   Levitt, M.H., Raleigh, D.P., Creuzet, F., Griffin, R.G., J.Chem.Phys. 92 (1990) 6347.
[28]   Eden, M. and Levitt, M.H., J. Chem. Phys. 111 (1999) 1511.
[29]   Brinkmann, A., Eden, M. and Levitt, M.H., J. Chem. Phys. 112 (2000) 8539.
[30]   Carravetta, M., Eden, M., Zhao, X., Brinkmann, A. and Levitt, M.H., Chem.Phys.Lett. 321 (2000), 205.
[31]   Carravetta, M., Eden, E., Levitt, M.H., manuscript in preparation
[32]   Wokaun, A., Ernst, R.R., Chem.Phys.Lett. 52 (1977) 407.
[33]   Maricq, M.M. and Waugh, J.S., J.Chem.Phys. 70 (1979) 3300.
[34]   Antzutkin, O.N. and Levitt, M.H., J.Magn.Reson. 118 (1996) 295.
[35]   Haberkorn, R.A., Stark, R.E, van Willigen, II., Griffin, R.G., J.Am.Chem.Soc. 103 (1981) 2534.
[36]   Jönsson, P.G., Kvick, A., Acta.Cryst. B28 (1972) 1827.
[37]   Hexem, J.G., Frey, M.H., Opella, S.J., J.Am.Chem.Soc. 103 (1981) 226.
[38]   Olivieri, A.C., J.Magn.Reson. 81 (1989) 201.

Solid state ^{19}F-NMR of biomembranes

Stephan L. Grage, Jesús Salgado[a], Ulrich Dürr, Sergii Afonin, Ralf W. Glaser and Anne S. Ulrich*

*Institute of Molecular Biology, Friedrich-Schiller-University of Jena, Winzerlaer Str. 10, 07745 Jena, Germany, *email: ulrich@molebio.uni-jena.de*
[a] *New address: Universitat de Valencia, Dep. Bioquímica i Bilogía Molecular, Dr. Moliner, 50, 46100 Burjassot (Valencia), Spain*

Introduction

Structure determination of membrane-associated polypeptides presents one of the major challenges to solid state NMR spectroscopy. Many studies have been carried out so far using selective isotope labels, such as ^{2}H, ^{13}C, or ^{15}N[1]. These NMR-reporters can be incorporated into the protein backbone or side chains, to reveal local structural parameters and to describe the dynamic properties of the membrane-embedded molecule. For example, an distance r can be measured between a pair of labels by means of dipolar recoupling MAS techniques such as rotational resonance or REDOR[2]. Alternatively, uniaxially oriented samples are used to determine the angle θ of a labelled molecular segment with respect to the membrane normal N[3,4,5]. The latter approach relies on the orientation-dependent resonance frequency, which carries information about the anisotropic chemical shift tensor (^{13}C, ^{15}N), the dipolar

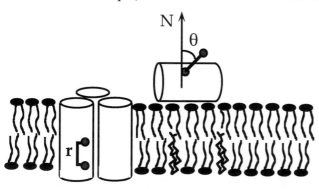

In membrane-associated polypeptides it is possible to determine local structural parameters, such as an internuclear distance r or an orientational angle θ, by solid state NMR of selectively labelled samples.

S.R.Kiihne and H.J.M.deGroot (eds.), Perspectives on Solid State NMR in Biology, 83-91.
©2001 Kluwer Academic Publishers. Printed in the Netherlands.

coupling (^{1}H-^{15}N), or the quadrupolar interaction (^{2}H)[6,7].

    Given the intrinsic two-dimensional nature of the lipid bilayer, it is rather straightforward to prepare macroscopically oriented samples of membrane proteins in any desired lipid environment on small glass plates. These kinds of samples, where a stack of about 15 plates carries up to 50 thousand of bilayers, will be used in all of the applications discussed below.

    Since conventional isotope labels typically exhibit low sensitivity, we employ fluorine as an alternative reporter nucleus; it is orders of magnitude more sensitive and may serve as a background-free NMR probe in biological compounds[8]. The broad chemical shift anisotropy of ^{19}F is highly informative about orientational constraints, and its strong dipolar interactions are advantageous when long range distance measurements are required, e.g. for monitoring helix-helix interactions.

    Compared to other biophysical labels that are commonly applied for structure analysis, such as fluorescence-probes or ESR spin labels, it is known that selective ^{19}F-labels cause much less perturbation of the native molecular structure. Several ^{19}F-labelled amino acids can be readily incorporated biosynthetically.

	Solid-state NMR		ESR	Fluorescence
	^{2}H, ^{13}C, ^{15}N	^{19}F		
Size increase	1.0	1.3	25	50
Typical sensitivity	10 μmol 10 hours	10 μmol 10 min	100 nmol 1 min	1 nmol 1 sec
Max. distance	7 Å	16 Å	50 Å	100 Å
Biosynthesis	+	±	−	−

Various labels can be used for structural studies

Methods

To facilitate structural studies of polypeptides in their native lipid environment, we have developed several different solid state ^{19}F-NMR techniques, all of which make use of uniaxially oriented membrane samples. The most simple approach to obtain an orientational constraint is to analyze the anisotropic ^{19}F chemical shift of a single labelled segment. Commonly, (near-)axially symmetric tensors are encountered for the ^{15}N CSA, the quadrupolar interaction, or dipolar couplings. In contrast, the CSA tensors of most ^{19}F-substitutents are highly non-axially symmetric and therefore reveal the 3-dimensional orientation of the labeled segment rather than the alignment of a single axis only. This valuable information, however, needs to be extracted by analyzing a full set of ^{19}F-NMR lineshapes from several different sample tilt angles.

The distance r between two ^{19}F-labels can be determined from the homonuclear dipolar coupling $\Delta\nu_{FF}(\theta)$, for example by measuring the splitting in the powder spectrum of a static non-oriented sample. The dipolar coupling also carries information about the

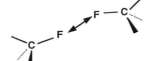 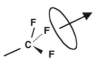

CSA-tensor:	*dipolar coupling:*	*2-spin system:*	*3-spin system:*
alignment in the membrane	order parameter and orientation	distance and internuclear angle	orientation of CF$_3$-group

angle θ between the internuclear vector and the membrane normal, provided that an oriented sample can be measured at a horizontal alignment in the external magnetic field.

$$\Delta\nu_{FF}(\theta) = \frac{3}{2}\left(\frac{\gamma_F^2\hbar}{\pi r^3}\right)\frac{(3\cos^2\theta-1)}{2} \qquad \text{Equation 1}$$

Since oriented samples are intrinsically incompatible with the conventional MAS techniques that are used to measure dipolar couplings, we have employed static multipulse techniques to separate the homonuclear dipolar interaction from other sources of linebroadening (i.e. chemical shift anisotropy, ^{1}H-^{19}F heteronuclear coupling, magnetic field inhomogeneities). An improved CPMG sequence with composite pulses and xy8 phase cycling has successfully yielded purely dipolar spectra of various doubly-labelled compounds incorporated in lipid bilayers[9]. By analyzing the orientation-dependence of the dipolar interaction, it is possible to determine both the internuclear angle θ as well as the distance r between two labelled sites according to equation 1. In the case of small molecules that are highly mobile in the liquid crystalline phase of the lipid bilayer, the distance information is replaced by an order parameter S$_{mol}$. The order parameter is defined as the time averaged orientation of the molecular axis, <3cos^{2}θ-1>/2, thus describing the degree of motional averaging.

Results and Discussion

After an initial validation of the ^{19}F-NMR CPMG approach with various doubly-labelled lipids, we extended our studies to a membrane-soluble drug, flufenamic acid, which carries a CF$_3$-group[10]. Whereas the dipolar coupling between two spins

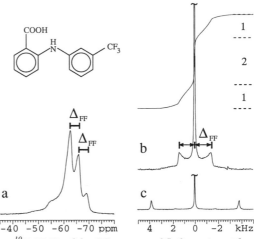

^{19}F-NMR of the CF$_3$-group of flufenamic acid

gives rise to a doublet, the 3-spin system of the CF_3-group produces a triplet. The dipolar splitting $\Delta\nu_{FF}(90°)$ is already apparent from the standard ^{1}H-decoupled powder spectrum of flufenamic acid embedded in a DMPC dispersion (a), but the dipolar lineshape is still convoluted with the broad CSA. The CPMG experiment, on the other hand, reveals the pure dipolar powder spectrum (b), notably without any need for ^{1}H-decoupling. The integrated intensities show the expected ratio of 1:2:1. Further line-narrowing is achieved by preparing an oriented sample (c), which yields a well-resolved splitting $\Delta\nu_{FF}(0°)$ and linewidths of several tens of Hz. The orientation-dependent dipolar splitting can now be used to describe the alignment of the CF_3-labelled molecule in the membrane. In the liquid crystalline phase the ^{19}F-NMR data shows that the motional axis of flufenamate is aligned with the bilayer normal, as expected. However, under conditions of low hydration we observed that the axis of motional averaging is flipped into the plane of the oriented sample. This analysis was based on an aligned sample measured at different tilt angles in the magnetic field. The same behaviour was observed by ^{31}P-NMR for a fraction of the phospholipids, suggesting that both the drug and some of the lipids assume a non-trivial morphology in the flat sample. We proposed that flufenamic acid can induce a hexagonal phase in DMPC, which is rather uncommon for this type of phospholipid, but which was confirmed by electron microscopy[11].

We also studied the dipolar couplings between two CF_3-groups. Due to the enormous resolution enhancement of the multipulse narrowing applied to oriented samples, it was possible to resolve the complex dipolar pattern of a 6-spin system consisting of more than 20 lines[Grage et al., in preparation].

To make ^{19}F-NMR fully available for structural studies of polypeptides, we have measured systematically the CSA tensors and relaxation properties of various aliphatic and aromatic ^{19}F-labelled amino acids over a broad range of temperatures. [Dürr et al., in preparation] The polycrystalline powders show a variety of asymmetric powder lineshapes, and the components of their chemical shift tensors were extracted from MAS spectra using the Herzfeld-Berger algorithm[12]. Knowledge of the principal CSA tensor values is essential for an interpretation of the orientation and dynamic behaviour of all further ^{19}F-labelled biological compounds.

Having established the basic ^{19}F-NMR experimental techniques with oriented samples, and having compiled a data base of ^{19}F-NMR parameters for fluorine-labelled amino acids, these tools are now being applied to polypeptides in membranes. In order to assess the validity of the new approach, we first examined the geometries and mobilities of the labelled side chains in the rigid framework of gramicidin A, whose structure is well-known from ^{2}H-, ^{15}N-, and ^{13}C-NMR[13]. The antibiotic peptide (15 aa) forms a β-helical ion channel in membranes. The channel rim is lined with four Trp residues which play an important role in ion conductivity and selectivity. Hence, it was not only of methodological but also of biological interest to examine the chemical shift anisotropy of ^{19}F in the 5F-Trp side chains. When the labelled peptide is incorporated into DMPC membranes, the ^{19}F-NMR spectra respond to the phase

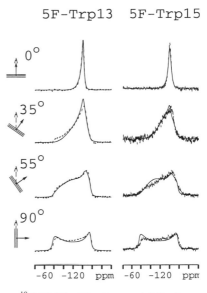

^{19}F-NMR of Trp in gramicidin A

behaviour of the lipid: axially symmetric powder spectra indicate that the whole peptide is rotating in the liquid crystalline phase, whereas the full-width non-axially symmetric powder spectra in the gel state showed that there is hardly any motional averaging. When an oriented sample was tilted in the magnetic field, it yielded a series of lineshapes which allowed the CSA tensor to be positioned into the known molecular frame of the 5F-Trp side chain[14]. This analysis reveals that σ_{22} points along the C-F bond, and σ_{33} lies perpendicular to the indole plane.

By ^{19}F-NMR the side chain torsion angles were analyzed in the gel state, and they can be compared with previous ^{2}H-NMR results on the same 5F-Trp labelled gramicidin A in the liquid crystalline phase[15]. The previous study had demonstrated that the ^{19}F-substituent perturbs the peptide torsion angles by up to 10°. Our comparison, however, yielded differences of up to 20°, illustrating the phase state of the lipid exerts a more significant influence on the side chain orientation and dynamics. Based on this first characterization of the ^{19}F CSA tensor of 5F-Trp, it will now be possible to address the side chain conformations in other proteins, which can often be labelled biosynthetically with fluorinated Trp. Notably, this particular side chain is frequently found in α-helices near the hydrophilic-hydrophobic bilayer interface.

To fully describe the alignment and mobility of a polypeptide in a membrane by ^{19}F-NMR, however, it is essential to have a handle on the backbone rather than examining a flexible side chain. Several amino acids are highlighted in the figure on the previous page as rigid probes of the polypeptide backbone, and the following examples will focus on the non-natural 4F-phenylglycine (4F-Phg). This particular side chain is well suited to determine the orientation of any known secondary structure element (α-helix, β-sheet) within the membrane because it protrudes stiffly from the peptide backbone. Using automated peptide synthesis, 4F-Phg is readily substituted for any other residue, most preferably a bulky hydrophobic one. Although we encountered a strong tendency of this side chain to racemize, the resulting D- and L- peptides could usually be separated by HPLC. The two racemic forms of 4F-Phg are readily identified by Marfey's reaction after acid hydrolysis of the peptide[16].

Our first application of ^{19}F-NMR to a 4F-Phg labelled peptide was aimed at describing the interaction of gramicidin S with membranes. This antimicrobial peptide with the sequence cyclic-(Val-Orn-Leu-DPhe-Pro)$_2$ forms an antiparallel β-sheet structure with an amphipathic character, which gives it a high affinity for lipid bilayers. The peptide is able to destabilize bacterial membranes, possibly through the formation of pores, however the mechanism of action has not yet been explained at the molecular level[17]. A ^{19}F-labelled gramicidin S analogue (F-GS) was prepared, in which the two equivalent Leu residues are replaced by 4F-Phg. ^{19}F-NMR showed that the mobility of F-GS in DMPC model membranes is closely linked to the phase state of the lipid. At low temperature the ^{19}F-labels display a static tensor lineshape, which becomes axially averaged above the main phase transition.

The alignment of F-GS in membranes was determined by CSA analysis of the two 4F-Phg side chains in oriented samples[18]. At low peptide concentration, only a single resonance line is observed, indicating that both side chains are aligned at the same angle with respect to the sample normal. We conclude that the molecule is positioned symmetrically and flat in the bilayer, which is consistent with its amphiphilic character. The peptide undergoes fast axial rotation with an order parameter of S ≈ 0.36 in the liquid crystalline phase of the lipid.

A remarkable change in the alignment of F-GS was observed in the temperature range around the main phase transition of DMPC. In samples with high peptide:lipid ratios (≥1:40), two additional resonances appear and dominate the ^{19}F-NMR spectra. These signals are attributed to the two 4F-Phg residues, which must be oriented at different angles with respect to the membrane normal. From the respective resonance positions, the CSA analysis defines the alignment of the peptide backbone as *upright* within the lipid

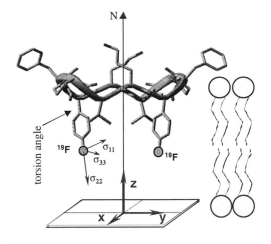

[19]*F-NMR of the antimicrobial peptide gramicidin S, labelled with 4F-phenylglycine, reveals that the amphiphilic structure is aligned symmetrically in the lipid bilayer at low peptide concentration.*

bilayer [Salgado et al., in preparation]. This result demonstrates that F-GS is able to form an oligomeric structure in the membrane, which may allow the passage of polar molecules and explain the permeabilization of membranes. Several different oligomeric pore models for gramicidin S, all of which can be stabilized by intermolecular H-bonds, are currently being evaluated in terms of the orientational constraints from ^{19}F-NMR. The most promising candidates are a β-barrel structure like "porin", or a stack of horizontally aligned rings. To determine the exact alignment of the peptide, additional input about the 4F-Phg torsion angles is required, which will be available from further F-GS analogues.

Another peptide that is currently being examined by ^{19}F-NMR in our lab is derived from the sea urchin fertilization protein "bindin", which mediates membrane fusion between sperm and egg. A fully conserved 18-residue amphipatic sequence "B18" represents the minimal functional part of the 24 kDa protein[19,20]. The structure of the B18 peptide was investigated by ^{1}H-NMR in several membrane-mimicking environments: in 30% TFE as well as in DPC or SDS micelles[21]. It forms two helical segments connected by a flexible loop. For solid state ^{19}F-NMR, nine different fluorinated B18 analogues, each with a single Leu to 4F-Phg substitution, were synthesized. In oriented samples, all B18 analogues exhibited single narrow ^{19}F-NMR resonances with different chemical shifts, which suggests a uniform peptide structure. [Afonin et al., in preparation] The chemical shift values contain information about the orientations of the individual 4F-Phg side chains, and their multiplicity helps to analyze the B18 structure even if the torsion angle of 4F-Phg should remain undetermined. For this analysis of membrane-bound B18, the same helix-turn-helix motif was assumed as determined by ^{1}H-NMR in solution[21]. The respective orientations of the two helical segments were described by two independent sets of Euler angles α and β, and all combinations of α and β were swept through to calculate the expected ^{19}F-NMR frequencies. By comparing these predictions with the experimental set of data, we obtained a family of structures that are consistent with a tilted alignment of the fusogenic peptide in the lipid bilayer[22].

Conclusions

Solid state ^{19}F-NMR of suitably labelled polypeptides in oriented membranes offers several distinct advantages over structural studies with conventional isotope labels, namely in terms of high sensitivity and accessibility of large distances. Well resolved ^{19}F-NMR signals can be observed in favourable cases for only 0.1 mg of peptide within less than one hour. The high sensitivity thus allows a comprehensive and systematic characterization of a membrane system over a wide range of conditions. Such measurements are essential for many membrane-associated peptides, whose alignment and oligomerization state are affected by changes in peptide:lipid ratio, temperature, or lipid composition.

The orientation and dynamics of labelled side chains can be readily determined from the ^{19}F chemical shift anisotropy in oriented samples, now that the fundamental

CSA parameters have been made available for many different amino acids. Structural information about the alignment of a peptide backbone, too, is accessible by using the non-natural amino acid 4F-phenylglycine. The most demanding step in the interpretation of the ^{19}F chemical shift data from 4F-Phg is the distinction between side chain torsion angle and backbone orientation. These ambiguities, due to the non-axial symmetry, can be solved by combining the results of several ^{19}F-labelled positions. Alternatively, 4CF$_3$-Phg could be employed, which behaves as an axially symmetric label. Since the dipolar coupling within a CF$_3$-group can be measured with high precison using the CPMG experiment, this particular side chain will be the focus of some of our further ^{19}F-NMR studies in biomembranes.

Acknowledgements

We thank Douglas Young for preparing several ^{19}F-labelled leucine derivatives; Tim Cross and Junfeng Wang for providing the 5F-Trp-labelled gramicidin A; and Les Kondejewski, Ronald McElhaney and Robert Hodges for synthesizing the 4F-Phg labelled gramicidin S. The authors gratefully acknowledge the financial support of the Deutsche Forschungsgemeinschaft in the SFB 197 (TP B13), as well as the EMBO foundation and the Thüringer Ministerium for the fellowships of J.S. and S.L.G.

References

[1] Watts, A., Ulrich, A.S., and Middleton, D.A., *Mol. Membr. Biol.* 12 (1995) 233
[2] Fu, R., and Cross T.A. *Annu. Rev. Biophys. Biomol. Struct.* 28 (1999) 235
[3] Ulrich, A.S., Wallat, I., Heyn, M.P., and Watts, A., *Nature Struct. Biol.* 2 (1995) 190
[4] Asakura, T., Minami, M., Shimada, R., Demura, M., Osanai, M., Fujito, T., Imanari, M. and Ulrich, A.S., *Macromol.* 30 (1997) 2429
[5] Kameda, T., Ohkawa, Y., Yoshizawa, K., Naito, J., Ulrich, A.S., and Asakura, T., *Macromol.* 32 (1999) 7166
[6] Ulrich, A.S. and Watts, A., *Solid State NMR.* 2 (1993) 21
[7] Ulrich, A.S. and Grage, S.L., In *Solid state NMR of polymers*, In I. Ando and T. Asakura (eds.) Elsevier, Amsterdam, 1998, p. 190
[8] Ulrich, A.S. In *Encyclopedia of Spectroscopy and Spectrometry* In Lindon, J., Tranter, G., and Holmes, J. (eds.) Academic Press, 2000, p. 813
[9] Grage, S.L. and Ulrich, A.S., *J. Magn. Res.* 138 (1999) 98
[10] Grage, S.L. and Ulrich, A.S., *J. Mag. Res.* 146 (2000) 81
[11] Grage, S.L., Gauger, D., Selle, C., Pohle, W., Richter, W., and Ulrich, A.S., *Phys. Chem. Chem. Phys.* 105 (2000) 149
[12] Herzfeld, J., and Berger, A.E., *J. Chem. Phys.* 73 (1980) 6021
[13] Ketchem, R.R., Roux, B., and Cross, T.A., *Structure* 5 (1997) 1655
[14] Grage, S.L., Wang, J., Cross, T.A., and Ulrich, A.S., *JACS* (2001) to be submitted
[15] Cotton, M., Tian, C., Busath, D.D., Shirt, R.B., and Cross, T.A., *Biochem.* 38 (1999) 9185
[16] Marfey, P., and Ottesen, M., *Carlsberg Res. Commun.* 49 (1984) 585, and 591

[17] Prenner, E.J., Lewis, R.N.A.H., and McElhaney, R.N., *Biochim. Biophys. Acta* 1462 (1999) 201

[18] Salgado, J., Grage, S.L., Kondejewski, L.H., McElhaney, R.N., Hodges, R.S., and Ulrich, A.S., *J. Biomol. NMR* (2001) submitted

[19] Ulrich, A.S., Otter, M., Glabe, C., and Hoekstra, D. *J. Biol. Chem.* 273 (1998) 16748

[20] Ulrich, A.S., Tichelaar, W., Förster, G., Zschörnig, O., Weinkauf, S., and Meyer, H., *Biophys. J.,* 77 (1999) 829

[21] Glaser, R.W., Grüne, M., Wandelt, C., and Ulrich, A.S., *Biochemistry* 38 (1999) 2560

[22] Binder, H., Arnold, K., Ulrich, A.S., and Zschörnig, O. *Biochim. Biophys. Acta.* 1468 (2000) 345.

SECTION III:

Important Technologies:

Computational Techniques

Membrane Protein Expression

MR Microscopy

Numerical simulations for experiment design and extraction of structural parameters in biological solid-state NMR spectroscopy

Mads Bak, Robert Schultz, and Niels Chr. Nielsen
Laboratory for Biomolecular NMR Spectroscopy, Department of Molecular and Structural Biology, Science Park, University of Aarhus, DK-8000 Aarhus C, Denmark

Introduction

Based on a tremendous technological progress, during the past decade it has been demonstrated that solid-state NMR is capable of providing very detailed information about the structure and dynamics of biological molecules in the solid phase. Using state-of-the-art methodology, it is now realistic to resolve, assign, and structurally interpret resonances from peptides/proteins with about 50 – 100 residues and to obtain local structure information for proteins an order of magnitude larger. With considerable room for future development (higher magnetic fields, stronger rf fields, higher spinning speeds, new isotope labeling and expression methods, and new multi-dimensional pulse sequences), these achievements open up exiting perspectives for the study of, e.g., membrane proteins, protein aggregates, and colloids. This is of great importance considering that, e.g., membrane proteins in one way or another are involved in most biological processes *and* simultaneously may be extremely difficult to characterize at atomic resolution using traditional structure determination methods.

A major challenge, and obstacle, for structure determination of proteins in the solid phase concerns the resolution and sensitivity of the NMR spectra. The resolution problem has motivated the design of dozens of pulse sequences which, through transfer of coherence between, e.g. ^{1}H, ^{13}C, and ^{15}N spin species and acquisition of multi-dimensional spectra, may improve the resolution to an extent which allows structure determination for large peptides. The sensitivity problem has motivated the use of high-field equipment, optimized coherence transfer elements, and more exotic signal enhancement methods based on optical and microwave nuclear polarization transfer. Traditionally, two experimental approaches have formed the basis for biological solid-state NMR, namely methods applying to rotating powder samples [1] and methods

S.R.Kiihne and H.J.M.deGroot (eds.), Perspectives on Solid State NMR in Biology, 95-109.

relying on uniaxially oriented samples [2], although hybrids between the two have also been proposed [3]. In the former case it is fundamental to design pulse sequences which under conditions of magic-angle spinning (MAS) allow efficient decoupling/recoupling of the nuclear spin interactions to achieve the desired structure information or mediate the desired coherence transfer. In the latter case, the aim is to design multi-dimensional experiments which, based on the orientation dependence of anisotropic nuclear spin interactions, provide information about the structure of the protein as well as its orientation relative to the bilayer normal.

The present demand for advanced multiple-pulse sequences providing sufficient spectral resolution and structural constraints (torsion angles, internuclear distances) imposes a tremendous need for efficient tools to engineer optimum experiments and to extract structure information from experimental spectra. While pulse sequence engineering typically involves appropriate combinations of intuition, analytical methods [4-7], and numerical optimization [8], spectral interpretation typically requires numerical simulations and iterative fitting of simulated to experimental spectra. Confronted with these needs we have developed a general simulation program for solid-state NMR (SIMPSON) which essentially works like a "computer spectrometer" with flexibility similar to the normal interface on a solid-state NMR spectrometer [9]. In this paper we demonstrate the utility of SIMPSON in the design and application of solid-state NMR for the study of biological problems and introduce novel features relevant for structure determination in this area.

Methods

The SIMPSON software package has been designed with the intention of making a flexible tool for fast numerical simulations in solid-state NMR while maintaining a transparent and relatively easy to use interface, even in cases of quite complicated experiments. This requires that the user can define the Hamiltonian and control the stepwise manipulation of the density operator throughout a multiple-pulse experiment without the need for excessive computer coding and mathematics at the level of Hamiltonians and propagators. Formulated in other words, it should be possible to activate pulses, delays, phase cycling, filtering, and data manipulation with similar flexibility as on a spectrometer while simultaneously using state-of-the-art procedures for fast calculation of powder spectra [10,11]. Most conveniently, all manipulations of the nuclear spin system should be controlled at a "scripting" level, allowing for fast and straightforward implementation of advanced experimental methods.

To fulfill these needs, SIMPSON is controlled via a Tcl [12] input file containing four sections

1. spinsys   2. par   3. pulseseq   4. main

enabling independent definition of the spin system (rf channels, nuclei, interactions etc.), external parameters (field strengths, crystallites, initial/detection operators,

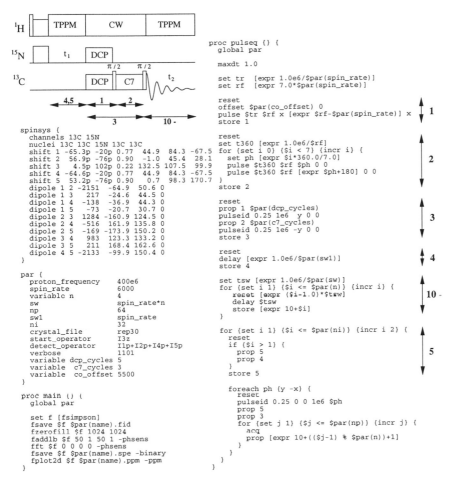

```
spinsys {
    channels 13C 15N
    nuclei 13C 13C 15N 13C 13C
    shift 1 -65.3p -20p 0.77   44.9   84.3 -67.5
    shift 2  56.9p -76p 0.90   -1.0   45.4  28.1
    shift 3   4.5p 102p 0.22  132.5  107.5  99.9
    shift 4 -64.6p -20p 0.77   44.9   84.3 -67.5
    shift 5  53.2p -76p 0.90    0.7   98.3 170.7
    dipole 1 2 -2151  -64.9  50.6 0
    dipole 1 3   217  -24.6  44.5 0
    dipole 1 4  -138  -36.9  44.3 0
    dipole 1 5   -73  -20.7  30.7 0
    dipole 2 3  1284 -160.9 124.5 0
    dipole 2 4  -516  161.9 135.8 0
    dipole 2 5  -169 -173.9 150.2 0
    dipole 3 4   983  123.3 133.2 0
    dipole 3 5   211  168.4 162.6 0
    dipole 4 5 -2133  -99.9 150.4 0
}

par {
    proton_frequency      400e6
    spin_rate             6000
    variable n            4
    sw                    spin_rate*n
    np                    64
    sw1                   spin_rate
    ni                    32
    crystal_file          rep30
    start_operator        I3z
    detect_operator       I1p+I2p+I4p+I5p
    verbose               1101
    variable dcp_cycles   5
    variable c7_cycles    3
    variable co_offset    5500
}

proc main {} {
    global par

    set f [fsimpson]
    fsave $f $par(name).fid
    fzerofill $f 1024 1024
    faddlb $f 50 1 50 1 -phsens
    fft $f 0 0 0 0 -phsens
    fsave $f $par(name).spe -binary
    fplot2d $f $par(name).ppm -ppm
}
```

```
proc pulseq {} {
    global par

    maxdt 1.0

    set tr  [expr 1.0e6/$par(spin_rate)]
    set rf  [expr 7.0*$par(spin_rate)]

    reset
    offset $par(co_offset) 0
    pulse $tr $rf x [expr $rf-$par(spin_rate)] x
    store 1

    reset
    set t360 [expr 1.0e6/$rf]
    for {set i 0} {$i < 7} {incr i} {
        set ph [expr $i*360.0/7.0]
        pulse $t360 $rf $ph 0 0
        pulse $t360 $rf [expr $ph+180] 0 0
    }
    store 2

    reset
    prop 1 $par(dcp_cycles)
    pulseid 0.25 1e6 y 0 0
    prop 2 $par(c7_cycles)
    pulseid 0.25 1e6 -y 0 0
    store 3

    reset
    delay [expr 1.0e6/$par(sw1)]
    store 4

    set tsw [expr 1.0e6/$par(sw)]
    for {set i 1} {$i <= $par(n)} {incr i} {
        reset [expr ($i-1.0)*$tsw]
        delay $tsw
        store [expr 10+$i]
    }

    for {set i 1} {$i <= $par(ni)} {incr i 2} {
        reset
        if {$i > 1} {
            prop 5
            prop 4
        }
        store 5
    }

    foreach ph {y -x} {
        reset
        pulseid 0.25 0 0 1e6 $ph
        prop 5
        prop 3
        for {set j 1} {$j <= $par(np)} {incr j} {
            acq
            prop [expr 10+(($j-1) % $par(n))+1]
        }
    }
}
```

Fig. 1. SIMPSON Tcl input file demonstrating the content of the spinsys, par, pulseseq, *and* main *elements for a 2D N(CO)CA pulse sequence based on DCP ^{15}N-^{13}C and C7 ^{13}C-^{13}C elements. The numbers in the pulse sequence (top) refer to the precalculated propators marked in the* pulseseq *code. Specifically the simulation involves a typical backbone $^{13}C_{\alpha,i-1}$-$^{13}C'_{i-1}$-$^{15}N_i$-$^{13}C_{\alpha,i}$-$^{13}C'_i$ five-spin system (extracted from the Gln^2 and Ile^3 residues of the ubiquitin 1D3Z [16] pdb file [17] using pdb2SIMPSON) under the assumption of ideal ^1H to ^{15}N transfer and ^1H decoupling. We note that 120 ppm has been subtracted from all isotropic chemical shifts (second entry in the* spinsys *shift lines [9]). This input file was used for one of the spin systems contributing to the simulation in fig. 4.*

sampling conditions etc.), the pulse sequence, and control of the simulation and data processing, respectively. To give an example of the content of these elements of the input file and to demonstrate that even quite complicated pulse sequences are relatively simple to program, fig. 1 gives the Tcl input file for a novel DCP [13] and C7 [14] based 2D N(CO)CA pulse sequence which in the specific simulation operates on a $^{13}C_\alpha$-$^{13}C'$-^{15}N-$^{13}C_\alpha$-$^{13}C'$ five-spin system. Briefly the input file defines the relevant anisotropic chemical shift and dipolar coupling interactions in the `spinsys` section, while the physical conditions for the experiment and definition of various general parameters is accomplished in the `par` section. The `main` section describes the flow of the simulation, including calculation (fsimpson) as well as saving, Fourier trans-formation, and plotting of the spectrum. Finally, the `pulseseq` section defines the pulse sequence with propagators calculated a priori for the DCP part (marked 1), C7 part (marked 2), the combined DCP-C7 mixing sequence including the bracketing pulses (marked 3), the t_1 evolution (marked 4, 5), and the sampling period (marked 10 -). These propagators are used in the calculation of two 90° phase shifted 2D free-induction decays to obtain pure absorption spectra [15]. A detailed description of the various commands can be found in Ref. 9. In relation to the simulation, it is relevant to mention that the SIMPSON software package contains various productivity tools SIMPLOT, SIMDPS, and SIMFID for interactive plotting, pulse sequence visualization, and external data manipulations, respectively. The SIMPSON package is freely available as Open Source Software for the most common operating systems including Linux, Windows, and Unix [9]. The simulations in this paper were conducted on a standard Linux-controlled 450 MHz Pentium III processor.

Results and Discussion

As demonstrated in Ref. 9, the SIMPSON package represents a powerful tool for simulation of virtually all solid-state NMR experiments as implemented on the spectrometer. This offers interesting possibilities in relation to biological solid-state NMR. First, it allows simulation and direct comparison of state-of-the-art methods which by rf manipulations tailor the Hamiltonian to selectively accomplish certain coherence transfers or evolution under specific parts of the Hamiltonian. This is quite important since the scaling of the effects from undesired (i.e., suppression) as well as desired (i.e., maximization) components may depend strongly on the spin system and the size of the various anisotropic interactions and thus may differ considerably from application to application. Second, it allows straightforward investigation of the impact from error terms such as rf inhomogeneity and small (often hidden) delays for the physical accomplishment of amplitude, phase, and frequency shifts. Third, based on empirical or theoretical (ab initio, density functional theory) parameters for the various nuclear spin interactions, it is possible to systematically investigate pulse sequences for assignment and structure determination prior to experiments on the spectrometer. Fourth, based on numerical simulations it is possible to iteratively fit experimental spectra to extract information about the anisotropic interaction tensors bearing detailed information about the atomic and electronic structure of the molecule.

In the following, we will demonstrate some of these aspects in relation to powder as well as uniaxially oriented samples. We will demonstrate how SIMPSON in combination with a pdb-file interpreter and 3D GeomView graphics [18] may be setup

Table 1. Typical parameters for the peptide plane chemical shift tensors[a]

Spin	δ_{iso}	δ_{aniso}	η	α_{PE}	β_{PE}	γ_{PE}	Reference
^1H	9	-9	0.42	0	-90	0	20,21
$^{13}C_\alpha$	50	-20	0.43	90	90	0	22
$^{13}C'$	170	-76	0.90	0	0	95	23-25
^{15}N	119	99	0.19	-90	-90	-17	21,24,25

[a]*Chemical shifts in ppm relative to TMS (^1H, ^{13}C) and NH$_3$ (^{15}N) and Euler angles in degrees. Definitions of the parameters can be found in Ref. 9.*
[b]*We note that the orientation of the $^{13}C_\alpha$ chemical shift tensor may vary significantly depending on the residue type and secondary structure [26].*

to meet many of the needs presently existing for numerical simulations in biomolecular solid-state NMR. We note that the major aim is not to present new methods but rather to demonstrate how various methods may be systematically investigated/compared and may be seen as elements in an evolutionary process towards powerful experimental techniques.

A pdb to SIMPSON converter: Accurate numerical simulation of solid-state NMR spectra for polypeptides typically requires knowledge to the magnitude and orientation of several anisotropic nuclear spin interaction tensors involved in the evolution of the density operator during the pulse sequence. Obviously, it may be the objective of the experiment to determine one or more of these parameters while other parameters are fixed to realistic values to avoid too many variables in an optimization problem. In other cases the aim of the numerical simulation is to investigate the performance of an experiment on a realistic spin system with typical parameters. In all cases, it may be tedious manually to establish parameter sets for simulations involving a large number of spins.

For polypeptides, a reasonable set of parameters may be established considering that most of the chemical shifts tensors to a good approximation may be described by a unified set of parameters describing the magnitude of the tensor and its orientation relative to the peptide plane [19]. Likewise, the dipolar coupling tensors are axially symmetric, with a coupling constant of $b_{IS}/2\pi = -\gamma_I\gamma_S\hbar\mu_0/(4\pi r_{IS}^3)$, and oriented along the internuclear axis. Table 1 gives typical parameters for the magnitude (δ_{iso}, δ_{aniso}, η) and orientation (α_{PE}, β_{PE}, γ_{PE}) of the ^1H, $^{13}C_\alpha$, $^{13}C'$, and ^{15}N chemical shift tensors relative to the peptide plane as visualized in fig. 2A using GeomView 3D graphics. The magnitude is visualized by a standard tensor ellipsoid with X, Y, and Z labeling the δ_{xx}, δ_{yy}, and δ_{zz} elements for each tensor in its principal axis frame (P). The Euler angles relate P to the peptide frame (E), defined with the x axis along the N-H bond and the z axis perpendicular to the peptide plane.

Equipped with this information, it is possible to establish typical values for the chemical shift and dipolar coupling interaction tensors directly from *a priori* knowledge

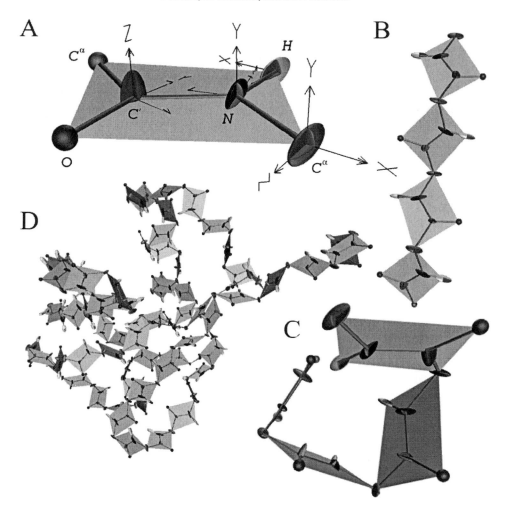

Fig. 2. Graphical representation of the pdb2SIMPSON tool for visualization of chemical shift and dipolar coupling tensors and establishment of their magnitude and orientations for the SIMPSON input file on the basis of structures in the pbd format. (A) Peptide plane, (B,C) four peptide planes of (B) β-pleated sheet (φ=-120°, ψ=120°) and (C) α-helix (φ=-65°, ψ=-40°) structures, and (D) the liquid-state NMR structure of ubiquitin [16] (pdb i.d. 1D3Z).

of the structure of the molecule that the spin system is associated with. Accordingly, we have constructed a program pdb2SIMPSON, which reads in a pdb file [17] containing the atomic coordinates for a full structure or for a structure element, specification of the relevant nuclear spin species, and potential limitations on the dipolar coupling constants. The output is the `spinsys` section of a SIMPSON input file (see fig. 1) and a GeomView 3D representation of the structure. This tool renders it quite easy to explore the performance of different pulse sequences on different structures such as β-pleated sheets (fig. 2B) and α-helices (fig. 2C) using SIMPSON. The structure elements may be established from a pdb file for a larger structure, such as ubiquitin (fig. 2D), known from XRD or liquid-state NMR or constructed by modeling (e.g., using WHAT IF [27]) or molecular dynamics simulations. As shall be demonstrated in the last part of the paper, this provides the possibility of testing pulse sequences not only on ideal structures, but also on "real" secondary structure elements and full structures possessing significant variations in the torsion angles.

b. SIMPSON for powder samples: A first stage in a typical strategy for the design of new solid-state NMR techniques involves the use of approximate analytical tools, such as average Hamiltonian theory [4,5], to establish pulse sequence elements suppressing or emphasizing certain nuclear spin interactions to as high an order as possible. In a subsequent stage, the methods are verified by experiments on appropriate models, which again may be verified by exact numerical simulations. We note that numerical simulations, obviously, could also be involved at the initial development stage, e.g., using SIMPSON to evaluate the manipulation of the internal Hamiltonian by external interactions either by exact calculation or by evaluation of average Hamiltonians.

The SIMPSON software package, in combination with pdb2SIMPSON, proves very useful for numerical simulations for pulse sequence engineering through numerical optimization and for testing of new experimental methods prior to experimental verification. Furthermore, using the same setup, it is straightforward to numerically simulate and compare the performance of different pulse sequences that may be used for the same purpose. As an example of the latter aspect, it may be of interest to numerically evaluate various approaches using combinations of ^{15}N to ^{13}C and ^{13}C to ^{13}C coherence transfers to establish intra- and inter-residue backbone and side-chain correlations for assignment purposes [28,29]. Addressing the peptide backbone fragment in fig. 3A, a typical experiment initially transfers ^{1}H coherence to ^{15}N coherence which after chemical shift encoding during the t_1 period is transferred to $^{13}C_\alpha$ in the same residue, $^{13}C'$ in the preceding residue, or to both depending on the desired ^{13}C-^{15}N shift correlation. In this context, it is particularly important to construct the pulse sequence from building blocks which accomplish selective transfer between the relevant spins *while* simultaneously preventing leakage of coherence to other spins. This requires evaluation and selection of building blocks that typically have been designed on spin-pair principles, for their performance in multiple-spin systems.

With these aspects in mind, figs. 3B and 3C show calculated efficiencies for ^{13}C to ^{15}N single-quantum transfer using RFDRCP [30], DCP [13], SPICP [31], C7 [14], and POST-C7 [32] with specific attention to their performance in N(CO)CA (fig. 3B) and N(CA)CB (fig. 3C) experiments. From these plots (optimized for transfer to the desired ^{13}C in $^{13}C'$-^{15}N-$^{13}C_\alpha$ three-spin systems), it becomes evident that DCP and C7 allow almost perfect truncation of the spin dynamics to the relevant spin-pair system

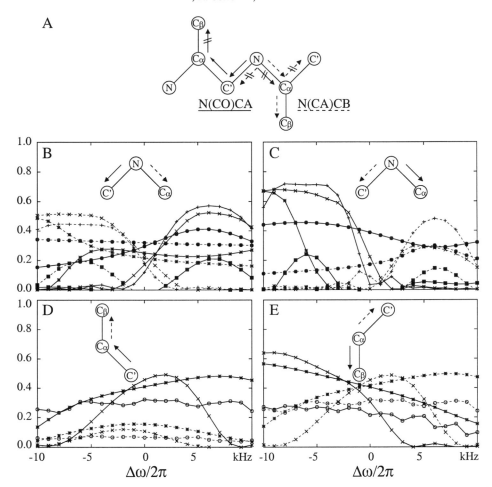

Fig. 3. Efficiencies for coherence transfers relevant for peptide backbone and side-chain assignments. (A) Peptide backbone fragment indicating transfers relevant for N(CO)CA and N(CA)CB experiments (undesired transfers indicated by crossed arrows). (B,C) ^{15}N to ^{13}C transfer in $^{13}C'$-^{15}N-$^{13}C_\alpha$ spin systems optimized for (B) ^{15}N to $^{13}C'$ and (C) ^{15}N to $^{13}C_\alpha$ transfer. (D,E) ^{13}C to ^{13}C transfer in $^{13}C'$-$^{13}C_\alpha$-$^{13}C_\beta$ spin systems optimized for (D) $^{13}C'$ to $^{13}C_\alpha$ and (E) $^{13}C_\alpha$ to $^{13}C_\beta$ transfer. All efficiencies have been calculated as function of the ^{13}C carrier offset ($\Delta\omega$) (referenced to the mean frequency between $^{13}C'$ and $^{13}C_\alpha$ carbons) using tensor values applying for $\omega_0/2\pi(^1H)= 400$ MHz, tensor orientations established by pdb2SIMPSON, $\omega_r/2\pi = 6$ kHz, and desired/undesired transfers indicated by solid/dashed lines (cf. diagrams). The rf field strengths (upper limit 42 kHz) and mixing times were optimized to give the most efficient transfer. Symbols: +: DCP; ×: C7; *: POST-C7; •: RFDRCP, ■: SPICP, and o: RFDR.

and still provide the efficiency of an ideal γ-encoded experiment [33]. In contrast, broadband methods, such as RFDRCP and POST-C7, cause leakage to the wrong spins leading to lower sensitivity and potentially undesired correlations in the 2D experiment unless purged later in the sequence. Following similar principles, figs. 3D and 3E show efficiencies for C7, POST-C7, and RFDR [34] (we note that DRAWS [35] performs very similar to RFDR) transfers between the various carbons in a $^{13}C'$-$^{13}C_\alpha$-$^{13}C_\beta$ three-spin system with the initial coherence at $^{13}C'$ (fig. 3D) and $^{13}C_\alpha$ (fig. 3E) with attention to N(CO)CA and N(CA)CB, respectively. In this case C7 is clearly the best experiment for selective transfer to the desired spin. This intuitively surprising result may easily be rationalized from the fact that C7 is broad-band with respect to chemical shift offsets as long as the carrier is close to the mean isotropic chemical shift of the two spins, i.e. $\Delta\omega_{C1} \approx -\Delta\omega_{C2}$. In contrast C7 is quite narrow-banded when $\Delta\omega_{C1} \approx \Delta\omega_{C2}$ implying, for example, that C7 irradiation at the mean frequency between $^{13}C_\alpha$ and $^{13}C'$ will provide efficient $^{13}C'$ to $^{13}C_\alpha$ transfer largely without leak to $^{13}C_\beta$ via $^{13}C_\alpha$ to $^{13}C_\beta$ transfer. Finally, we should note that a stringent optimization of the various correlation experiments would require several other techniques to be considered, being outside the scope of this demonstration of SIMPSON in evaluations of this type. Nonetheless, the importance of such evaluations is quite clear considering that experiments only providing 30% efficiency for each of the three transfer steps (including ^1H to ^{15}N transfer) will be associated with only 2.7%, 7.2%, and 21.6% of the efficiency of experiments giving 100%, 72%, and 50% efficiency for each step, respectively. We note that adiabatic transfers may under favorable conditions provide up to 100% efficiency [36,37], γ-encoded transfers up to 72% [33], and non-γ-encoded methods up to about 50%.

With appropriate control over the various coherence transfer schemes, a next step would naturally involve a performance test of the full pulse sequence. This may be conducted directly on the spectrometer or, as demonstrated in fig. 4, by a SIMPSON simulation. For the purpose of illustration we have chosen a $N_i(CO)C_{\alpha,i-1}$ type of experiment in our implementation using DCP for $^{15}N_i$ to $^{13}C'_{i-1}$ transfer and C7 for $^{13}C'_{i-1}$ to $^{13}C_{\alpha,i-1}$ transfer as represented by the pulse sequence and SIMPSON input file in fig. 1. The 2D spectrum in fig. 4 is calculated for the Gln^2–Val^5 fragment of ubiquitin using the liquid-state NMR isotropic chemical shifts from Wand et al. [38] and typical magnitudes and orientations of the anisotropic chemical shift and dipolar coupling tensors (cf. Table 1) associated to the known pdb structure using pdb2SIMPSON. The $^{15}NH(\omega_1/2\pi)$-$^{13}C_\alpha(\omega_2/2\pi)$ region of the spectrum demonstrates the selectively of the proposed 2D DCP-C7 N(CO)CA experiment by the clear dominance of the four cross-peaks corresponding to sequential (interresidue) correlations among the ubiquitin residues. It should be noted that the $^{13}C_\alpha$ resonance lineshapes are somewhat distorted by $n = 2$ rotational resonance effects [39] between the $^{13}C_\alpha$ and $^{13}C'$ resonances.

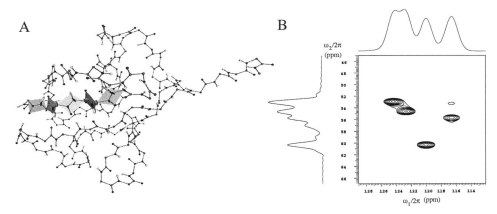

Fig. 4. N-C$_\alpha$ region of a 2D DCP-C7 N$_i$(CO)C$_{\alpha,i-1}$ spectrum (B) calculated for a powder of the Gln^2-Val^5 N-terminal residues of ubiquitin indicated in (A) using SIMPSON with anisotropic interaction tensors established from 1D3Z pdb structure [16] using pdb2SIMPSON and isotropic chemical shifts from Ref. 38. The spectrum was generated by superposition of spectra from four different ^{13}C$_\alpha$-^{13}C'-^{15}N-^{13}C$_\alpha$-^{13}C' spin systems. The experiment used $\omega_r/2\pi = \omega_{rf}/14\pi = 6$ kHz, 0.83 ms DCP mixing time, 1.0 ms C7 mixing time, 256 sampling points in both dimensions, and sampling of two spectra differing in the phase of the initial ^{15}N pulse to obtain phase sensitive 2D spectra (a typical input file is given in fig. 1). Each of the four spectra required about 4 h of CPU time on a cluster of four standard 450 MHz Pentium-III processors. The 2D spectrum was processed using SIMPSON and the Varian VNMR software.

d. SIMPSON for oriented samples: Obviously, SIMPSON is by no means restricted to powder samples. It may be applied equally well to simulate NMR experiments for liquid, single-crystal, and oriented samples. In this section we address the latter application with focus on the extraction of structural parameters and assignment of spectra for peptides incorporated into uniaxially oriented phospholipid bilayers.

It is well known that the orientation dependence of, e.g., the anisotropic ^{15}N chemical shift and ^{1}H-^{15}N dipolar coupling tensors may be exploited to determine the orientation of the peptide planes (E) relative to B_0 (L) as specified by the α_{EL} and β_{EL} Euler angles (fig. 5A). Thus, with the peptides incorporated into phospholipid bilayers oriented with the normal parallel to B_0, it is possible to determine not only the peptide structure but also the conformation of the peptide in the phospholipid membrane. These principles have formed the basis for a very important class of biological solid-state NMR experiments [2], among which the PISEMA experiment [40] has proven particularly useful for measurement of ^{15}N chemical shift and ^{1}H-^{15}N dipolar couplings for ^{15}N-labeled peptides. These parameters provide direct structural constraints in terms α_{EL}, β_{EL} angles as typically represented in so-called restriction plots illustrating the pairs of α_{EL}, β_{EL} angles compatible with the solid-state NMR data [41]. For the convenience of extracting structural parameters from this type of experiments, the SIMPSON package contains a module for direct interpretation of the observed chemical shift and

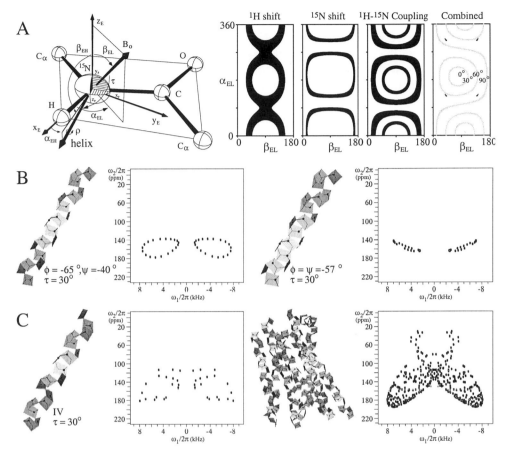

Fig. 5. (A) Typical restriction plots for ^1H chemical shift, ^{15}N chemical shift, and ^1H-^{15}N dipolar couplings generated by a module in the SIMPSON package. Specifically the plots corresponds to the Val17 residue in magainin2 with measured values of 9.9 ppm, 83.8 ppm, and 6.6 kHz for the ^1H shift, ^{15}N shift, and the ^1H-^{15}N dipolar coupling, respectively [44]. The dark areas indicate α_{EL}, β_{EL} orientational angles compatible with the measured values. The shaded restrictions correspond to ideal α-helices ($\phi=-65°$, $\psi=-40°$) tilted by the specified angles relative to B_0. (B,C) GeomView representations of (B) ideal α-helices (20 residues) with $\phi=-65°$, $\psi=-40°$ and $\phi=\psi=-57°$, (C) the IV helix of rhodopsin [45] (pdb i.d. 1F88 [17], oriented with a tilt of 30° relative to B_0) and all seven helices of rhodopsin (oriented with the average molecular axis along B_0) along with PISEMA spectra calculated using SIMPSON.

dipolar coupling parameters in terms of α_{EL}, β_{EL} restriction plots as demonstrated in fig. 5A.

Addressing uniaxially oriented samples, it may be relevant to investigate the applicability of different isotope labeling strategies and the use of state-of-the-art technology for uniformly labeled membrane proteins. This would allow several interesting questions to be raised and answered. For example, how ideal does an α-helix need to be to give socalled PISA wheels in PISEMA spectra [42,43] which may be assigned without the need for inter-residue correlations? How would the PISA wheels look for helices with typical spread in torsion angles and potentially kinked or bent structures? Which labeling scheme should be used if selective labeling were possible? These questions may be difficult and costly to answer experimentally, but may be addressed using SIMPSON using typical anisotropic interaction tensors (table 1) associated to a known atomic structure (pdb file) using pdb2SIMPSON. In fact, several of the questions may be answered by looking at the molecular structures and the corresponding simulated PISEMA spectra in figs. 5B and 5C. Comparison of the spectra for the two ideal α-helices with $\phi = -65°$, $\psi = -40°$ and $\phi = \psi = -57°$ (fig. 5B) reveals a noticeable dependence on the torsion angles, which in practice adds to other "wheel perturbing" fluctuations such as variations in the magnitude and orientation of the ^{15}N chemical shielding tensor as described by Cross an coworkers [43]. In this context, we should note that imperfect pulses (or insufficient decoupling) may represent another source of frequency shifts which is difficult to probe by analytical evaluations but is evaluated straightforwardly in a SIMPSON calculation. The significant dependence on the torsion angles may complicate automatic resonance assignment and direct extraction of helix tilt (τ) and rotational pitch (ρ) angles (see fig. 5A) from non-assigned experimental spectra. The latter becomes evident from the left-hand panels in fig. 5C showing a simulated PISEMA spectrum for a 30° tilted version of the most ideal α-helix (IV) of rhodopsin [45] which can only in part be interpreted as a PISA wheel. This situation is considerably worse for the other helices in rhodopsin as reflected in the simulated PISEMA spectrum for all seven transmembrane helices also shown in fig. 5C. Even without contributions from the loops, the spectrum has an almost powder-like character which may prevent interpretation in terms of wheel structures. Thus, in case it was possible to uniformly ^{15}N label 7TM proteins, such as rhodopsin in sufficient amounts, it would definitely be necessary to establish complementary structure information.

In non-ideal, albeit typical, cases as described above it may be necessary to assign the various PISEMA resonances either by experiments on selectively labeled samples (which may be difficult to obtain for larger proteins) or by support from inter-residue (e.g., ^{15}N-^{15}N) correlations [46]. Alternatively, relying on the validity of the empirical anisotropic interaction parameters and a reasonable model structure (e.g., from modeling, molecular dynamics calculations, liquid-state NMR, or XRD), it is possible to iteratively fit the experimental spectra for peptide fragments to establish both assignment and structural information. We have recently applied such iterative fitting procedures to obtain information about the orientation and structure of the alamethicin ionophore in lipid bilayers based on assigned (i.e., singly-labeled) ^{15}N chemical shift ^{1}H-^{15}N dipolar coupling correlated spectra [47]. In a further extended approach, the SIMPSON package contains programs which allow iterative fitting of assigned as well as unassigned PISEMA spectra to modeled structures. In the case of assigned resonances the optimization is straightforward, whereas the unassigned spectra represent a much more demanding task to be solved numerically. However, for relatively well-defined secondary structure elements such as the IV TM helix of rhodopsin (fig. 5C), the orientation of the proposed molecular model can be optimized to provide the highest degree of overlap between the experimental and modeled spectra. The best overlap can typically be established by minimizing the sum of distances from each resonance to the closest resonance in the experimental spectrum. Alternatively, given a small number of resonances, e.g., as obtained for selectively labeled peptides, all possible permutations of assignments can be searched for the best match. Under favorable conditions a similar approach may be used for larger proteins provided that different secondary structure elements can be identified, e.g., through their PISA wheel patterns in the experimental spectrum.

Conclusions

In conclusion, we have addressed the need for numerical simulations in modern biological solid-state NMR spectroscopy and introduced various tools which in combination with the SIMPSON simulation programs allow straightforward development and analysis of experimental methods for powder and uniaxially oriented samples. Obviously, most simulations may be conducted using custom-made software, as has been presented in the papers introducing the discussed techniques. It is important to realize, however, that this software is typically only available in the laboratories of the developers and may otherwise be difficult to use for the general NMR spectroscopist. Thus, the lack of appropriate software may hamper the dissemination of the most powerful techniques to the NMR application groups with the consequence of delayed scientific progress. Furthermore, it is relevant to emphasize the value of a "computer spectrometer" in the design of advanced solid-state NMR experiments both in the development process and also in the necessary performance tests and comparisons with state-of-the-art methodology. In this context it should be mentioned that most pulse sequence elements currently available have been developed on single-spin or spin-pair principles, which does not necessarily ensure optimum performance for multiple-spin systems. This particular point has been addressed in this paper by designing new 2D N(CO)CA and N(CA)CB pulse sequences based on structurally relevant multiple-spin systems. Likewise, with specific attention to oriented samples we have demonstrated that SIMPSON is useful as a means to investigate the impact of uniform isotope labeling on the NMR spectra and the potential of these for extraction of structural data. In addition to these specific examples, the SIMPSON software package may find important applications for students and researchers who want to investigate "wild ideas" that are too risky to motivate the synthesis of the right sample or the use of expensive spectrometer time, but upon success may contribute to the rapid development of biological solid-state NMR.

Acknowledgements

This research was part of projects supported from the European Commission (BIO4-CT97-2101), Carlsbergfondet, and the Danish Natural Science Council (J.No 9901954). We thank M.Sc. J.T. Rasmussen for important contributions in the initial stages of the SIMPSON project.

References

[1]   Griffin, R.G., Nature Struct. Biol. 5 (1998) 508.
[2]   Opella, S.J., Nature Struct. Biol. 4 (1997) 845.
[3]   Glaubitz, C. and Watts, A., J. Magn. Reson. 130 (1998) 305.
[4]   Haeberlen, U. and Waugh, J.S., Phys. Rev. 175 (1968) 453.
[5]   Hohwy, M. and Nielsen N.C., J. Chem. Phys. 109 (1998) 3780.
[6]   Kristensen, J., Bildsøe, H., Jakobsen, H.J. and Nielsen, N.C., Progr. NMR Spectrosc. 34 (1999) 1.
[7]   Untidt, T.S. , Schulte-Herbrüggen, T., Luy, B., Glaser, S.J., Griesinger, C., Sørensen, O.W. and Nielsen, N.C. Mol. Phys. 95 (1998) 787.
[8]   Untidt, T.S., Schulte-Herbrüggen, Sørensen, O.W. and Nielsen, N.C. J. Phys. Chem. A 103 (1999) 8921.
[9]   Bak, M., Rasmussen, J.T. and Nielsen, N.C., J. Magn. Reson. 147 (2000) 296. Open source software available from http://nmr.imsb.au.dk/.
[10]   Bak, M. and Nielsen, N.C., J. Magn. Reson. 125 (1997) 132.
[11]   Hohwy, M., Bildsøe, H., Jakobsen, H.J. and Nielsen, N.C., J. Magn. Reson. 136 (1999) 6.
[12]   Welch, B.B., Practical programming in Tcl and Tk, Prentice Hall PTR, New Jersey, 1995.
[13]   Schaefer, J., Stejskal, E.O., Garbow, J.R., and McKay, R.A., J. Magn. Reson. 59 (1984) 150.
[14]   Lee, Y., Kurur, N.D., Helmle, M., Johannesen, O.G., Nielsen, N.C. and Levitt, M.H., Chem. Phys. Lett. 242 (1995) 304.
[15]   States, D.J., Haberkorn, R.A., and Ruben, D.J., J. Magn. Reson. 48 (1982) 286.
[16]   Cornilescu, G., Marquardt, J.L., Ottiger, M., and Bax, A., J. Am. Chem. Soc. 120 (1998) 6836.
[17]   Protein Data Bank, http://www.rcsb.org/pdb/.
[18]   Levy, S., Munzner, T., Phillips, M. and others, Geomview 1.6.1, University of Minnesota, Minneapolis. 1996. Open source software. http://www.geomview.org/.
[19]   Cross, T.A. and Quine, J.R., Concepts Magn. Reson. 12 (2000) 55.
[20]   Gerald II, R., Bernhard, T., Haeberlen, U., Rendell, J. and Opella, S.J., J. Am. Chem. Soc. 115 (1993) 777.
[21]   Wu, C.H., Ramamoorthy, A., Gierasch, L.M. and Opella, S.J., J. Am. Chem. Soc. 117 (1995) 6148.
[22]   Naito, A., Ganapathy, S., Akasaka, K., and McDowell, C.A. J. Chem. Phys. 74 (1981) 3190.
[23]   Hartzell, C.J., Whitfield, M., Oas, T.G., and Drobny, G.P. J. Am.Chem.Soc. 109 (1987) 5966.
[24]   Teng, Q, Iqbal, M. and Cross, T.A., J. Am. Chem. Soc. 114 (1992) 5312.
[25]   Tan, W.M., Gu, Z., Zeri, A.C. and Opella, S.J., J. Biomol. NMR 13 (1999) 337.
[26]   Havlin, R.H., Le, H., Laws, D.D., deDios, A.C., and Oldfield, E., J. Am. Chem. Soc. 119

(1997) 11951.
[27]   Vriend, G., J. Mol. Graph. 8 (1990) 52.
[28]   Strauss, S.K., Bremi, T. and Ernst, R.R., J. Biomol. NMR 12 (1998) 39.
[29]   Hong, M., J. Biomol. NMR 15 (1999) 1.
[30]   Sun, B.Q., Costa, P.R. and Griffin, R.G., J. Magn. Reson. A 112 (1995) 191.
[31]   Wu, X. and Zilm, K., J. Magn. Reson. A 104 (1993) 154.
[32]   Hohwy, M.H., Jakobsen, H.J., Edén, M., Levitt, M.H. and Nielsen, N.C., J. Chem. Phys. 108 (1998) 2686.
[33]   Nielsen, N.C., Bildsøe, H., Jakobsen, H.J. and Levitt, M.H., J. Chem. Phys. 101 (1994) 1805.
[34]   Bennett, A.E., Ok, J.H., Griffin, R.G. and Vega, S., J. Chem. Phys. 96 (1992) 8624.
[35]   Gregory, D.M., Mitchell, D.J., Stringer, J.A., Kiihne, S., Shiels, J.C. Callahan, J., Mehta, M.A. and Drobny, G., Chem. Phys. Lett. 246 (1995) 654.
[36]   Baldus, M., Geurts, D.G., Hediger, S. and Meier, B.H., J. Magn. Reson. A 118 (1996) 140.
[37]   Verel, R., Baldus, M., Ernst, M. and Meier, B.H., Chem. Phys. Lett. 287 (1998) 421.
[38]   Wand, A.J., Urbauer, J.L., McEvoy, R.P. and Bieber, R.J. Biochemistry 35 (1996) 6116.
[39]   Raleigh, D.P., Levitt, M.H. and Griffin, R.G., Chem. Phys.Lett. 146 (1988) 71.
[40]   Wu, C., Ramamoorthy, A. and Opella, S.J. J. Magn. Reson. B109 (1994) 270.
[41]   Opella, S.J., Stewart, P.L. and Valentine, K.G. Quat. Rev. Biophys. 19 (1987) 7.
[42]   Marassi, F. and Opella, S.J., J. Magn. Reson. 144 (2000) 150.
[43]   Wang, J., Denny, J., Tian, C., Kim, S., Mo, Y., Kovacs, F., Song, Z., Nishimura, K., Gan, Z., Fu, R., Quine, J.R. and Cross, T.A., J. Magn. Reson. 144 (2000) 162.
[44]   Ramamoorthy, A., Marassi, F.M., Zasloff, M., and Opella, S.J., J. Biomol. NMR 6 (329) 1995.
[45]   Palczewski, K., Kumasaka, T., Hori, T., Behnke, C.A., Motoshima, H., Fox, B.A., Le Trong, I., Teller, D.C., Okada, T., Stenkamp, R.E., Yamamoto, M., and Miyano, M., Science 289 (2000) 739.
[46]   Marassi, F.M., Gesell, J.J., Valente, A.P., Kim, Y., Oblatt-Montal, M., Montal, M. and Opella, S.J., J. Biomol. NMR 14 (1999) 141.
[47]   Bak, M., Bywater, R.P., Hohwy, M., Thomsen, J.K., Adelhorst, K., Jakobsen, H.J., Sørensen, O.W. and Nielsen, N.C., Biophys. J. (submitted).

An ab-initio molecular dynamics modeling
of the primary photochemical event in vision

Francesco Buda[a], Sylvia I. E. Touw[b], and Huub J. M. de Groot[b]

[a] Department of Theoretical Chemistry, Vrije Universiteit Amsterdam,
De Boelelaan 1083, NL-1081 HV Amsterdam, The Netherlands
[b] Leiden Institute of Chemistry, Gorleaus Laboratories, Leiden University,
Einsteinweg 55, 2300 RA Leiden, The Netherlands

Introduction

In the process of vision, light stimuli are converted into neural information by a membrane protein called rhodopsin that initiates the transduction process. The primary photochemical event involves the photoconversion of rhodopsin into a metastable intermediate called bathorhodopsin [1,2]. The chromophore of rhodopsin is an 11-*cis* retinylidene prosthetic group covalently bound to the surrounding opsin protein via a protonated Schiff base (PSB) linkage to a specific lysine residue, Lys$_{296}$ (Fig. 1). The absorption of light induces the isomerization of the 11-*cis*-retinal chromophore to an all-*trans* configuration in bathorhodopsin. This photoprocess is completed in about 200 fs,

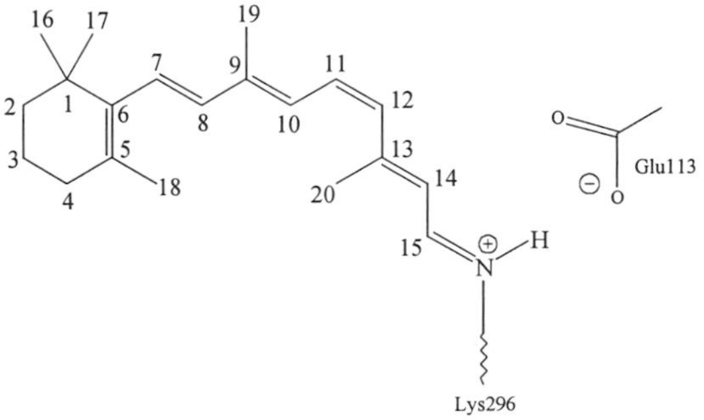

Fig.1:   Schematic structure of the 11-cis-retinylidene chromophore in bovine rhodopsin, showing the numbering system used.

S.R.Kiihne and H.J.M.deGroot (eds.), Perspectives on Solid State NMR in Biology, 111-122.

in the fastest photochemical reaction known [3], and is characterized by a high quantum yield of 0.67 [4]. The initial movements of the chromophore are thought to be tightly constrained by the surrounding protein, due to the very short time scale of photoisomerization [5]. This primary process stores an energy of about 32 kcal/mol [6] in bathorhodopsin and will then trigger slower conformational changes in the protein that initiate a chain of biochemical reactions eventually leading to the neural signal.

Very recently, a crystal   structure of rhodopsin has been determined from diffraction data with a moderate 2.8 Å resolution [7]. The configuration and structure of the retinylidene group is consistent with previously known spectroscopic information obtained with resonance Raman spectroscopy and nuclear magnetic resonance (NMR) [8-10]. For instance, at an early stage, solid-state NMR studies provided evidence for specific chromophore-counterion interactions in the region C12-C13 of the retinal [10]. This is confirmed by the X-ray model, which puts the negative counterion Glu_{113} close to the Schiff base nitrogen with distances between the carboxylate oxygen atoms and the N of 3.3 Å and 3.5 Å [7]. The resolution of the crystal structure is, as yet, not sufficient to measure accurately carbon-carbon distances and torsional angles within the chromophore. Recently, solid state magic angle spinning NMR has yielded high precision distance measurements in rhodopsin and in one of its photointermediates [11,12]. In the ground state, evidence was obtained for conformational out-of-plane distortions of a 12-s-*trans* retinal chromophore due to non-bonding steric intra-ligand and ligand-protein interactions. This distortion is generally considered crucial for an efficient isomerization process, but it has not yet been detected by X-ray.

Here we summarize some of the results obtained by means of *ab-initio* molecular dynamics simulations [13] on a model of the chromophore of rhodopsin that includes a number of high resolution structural restraints determined by NMR and the relevant interactions with the protein binding pocket. We discuss the ground-state geometry of the two configurations of the chromophore, before and after the photon absorption [14,15] and compare our results with available experimental data. In particular, we focus on the delocalization of the positive charge and how this affects the bond length alternation (BLA) along the conjugated carbon chain of retinal. We find that the energy is stored mainly in torsional strain in the primary photoproduct. By driving the system along the reaction coordinate with an external torsional field [16], we characterize the transition state in the isomerization process. During the relaxation from the transition state toward the all-*trans* photoproduct, we observe a charged, soliton-like defect which travels back and forth along the retinal on a time scale relevant for the isomerization.

Moreover, we present some preliminary results for a minimal model of the chromophore binding pocket based on the X-ray structure. This includes a more realistic counterion as determined by mutagenesis experiments [17] and NMR shift restraints [10], and confirmed by the recent crystal structure. We compute ^{13}C chemical shifts for this model and compare the results with the experiment and with recent Density Functional Theory calculations on retinylidene iminium salts [18].

Theoretical approach and description of the model

The *ab-initio* molecular dynamics approach describes the Newtonian dynamics of a system by using an interatomic potential obtained at each time-step from the instantaneous electronic ground state [13]. The electronic structure is obtained within the Density Functional Theory (DFT). We use a generalized gradient approximation for the exchange-correlation functional in the form proposed by Becke and Perdew [19,20]. Soft first-principles pseudopotentials [21] have been used to describe the interactions of the valence electrons with the inner cores. The Kohn-Sham single particle wavefunctions are expanded on a plane wave basis set with an energy cutoff of 20 Rydberg. The expansion in terms of plane waves implies the use of periodic boundary conditions. We have used a simulation box of 19 Å x 12 Å x 12 Å, large enough to avoid interaction with the images. The time step for the molecular dynamics simulations is 0.15 fs.

We have modeled the rhodopsin chromophore by an 11-*cis*-retinal PSB terminated with an ethylimine group to mimic the linkage of the chromophore to the protein lysine group. We have included in the initial configuration the structural constraints recently determined by magic angle spinning NMR [11]. In particular, the initial C20-C11 and C20-C10 distances have been set to the experimental values.

Solid-state NMR data have provided strong evidence that the side methyl groups bound to C9 and C13 experience steric interaction with protein residues, thus supporting the idea of a tight binding-pocket [5,11]. In our simulations, we have fixed in space the position of the carbon atoms of these methyl groups, to mimic the steric hindrance caused by the protein residues. These constraints are justified since essentially no rearrangement of the protein structure is expected during the short time (200 fs) of formation of bathorhodopsin. For the same reason we have fixed also the counterion and the carbon atom of the terminal methyl group.

The chromophore bears a net positive charge as a result of the protonation of the Schiff base (see fig. 1). We have modeled the counterion with a chloride ion (Cl⁻) placed at about 4.0 Å from C12. This counterion is much simpler than the carboxylate group in the glutamate residue Glu$_{113}$, however, at an earlier stage, semiempirical calculations showed that this counterion captures the essential electrostatic interaction, and yields a charge density distribution which is consistent with the experiment [22].

The chemical shift calculations on this model are performed within the Density Functional Theory framework using the technique described by Mauri et al. [23]. Norm-conserving pseudopotentials in the Kleinman-Bylander form have been employed [24]. The wavefunctions are expanded on a plane wave basis set with an energy cut-off of 70 Rydberg. Recently, this approach has been successfully applied to calculate the ^{13}C chemical shifts of retinylidene iminium salts [18].

To analyze a minimal model based on the X-ray structure with Density Functional Theory, we include an acetate ion to mimic the Glu$_{113}$ counterion, and we fix the positions of this group to the X-ray geometry. In addition, we fixed the positions of the carbon atoms of the methyl groups bound to the β-ionone ring to the experimental geometry. The methyl groups are clearly visible in the X-ray diffraction map and their positions are therefore assumed to be well-determined. The study of this model aims at exploring the relevance of a refined counterion geometry for the ground-state structure and chemical shifts of the retinylidene chromophore. The geometry optimization has been performed using the CPMD program package [25]. The Kohn-Sham single particle wavefunctions are expanded on a plane wave basis set with an energy cut-off of 25 Rydberg. The simulation box has been chosen 26 Å x 13 Å x 13 Å. The chemical shift calculations have been performed with Gaussian98 [26] within the Density Functional Theory framework, using the hybrid B3LYP exchange-correlation functional [27,28]. The calculations employed the Gauge Invariant Atomic Orbitals method [29] with the 6-31G basis set.

Results and Discussion

Ground State Structure and Isomerization: We have optimized the geometry of the the 11-*cis*-retinal PSB + Cl⁻ model with the geometrical constraints described in the previous section. The C20-C11 and C20-C10 distances after the optimization are 3.05 Å and 3.08 Å, respectively, in close agreement with NMR distances measurements [11]. The carbon-carbon bond lengths along the backbone chain are shown in Fig. 2. The 11-*cis*-retinal PSB presents clear bond alternation, except for the terminal region close to the nitrogen atom. For comparison, we have also shown the carbon-carbon distances for

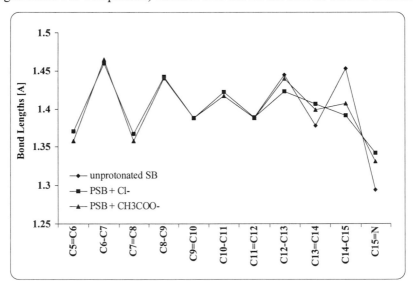

Fig. 2: Calculated bond length alternation pattern in the conjugated chain of the neutral 11-cis-retinal SB, the 11-cis-retinal PSB + Cl⁻, and the 11-cis-retinal PSB + CH₃COO⁻ model compounds.

an unprotonated Schiff base, which presents BLA along the entire backbone. Thus, the positive charge in the protonated chromophore induces a strong reduction of the BLA in the region close to the Schiff base nitrogen. The same effect has been observed in retinylidene iminium salts both experimentally [30] and theoretically [18]. This result indicates that the positive charge is delocalized over a few carbon-carbon bonds close to the nitrogen and that the single (double) bond character of these bonds is significantly altered. Specifically, the C14-C15 bond is shorter than the C13=C14 bond. The C14-C15 bond can therefore be assigned a partial double bond character, while the C13=C14 bond can be assigned a partial single bond character. In Fig. 2 we show also the result obtained for the model with the acetate counterion. It is clear that the delocalization of the positive charge is less pronounced than in the model with the chloride counterion. However, the BLA reduction is still evident, particularly in the C13-C14 and C14-C15 bonds. In this case, however, the C14-C15 bond is longer than the C13=C14 bond, and no bond order inversion takes place at C13.

The steric hindrance of the methyl group bound to C13 induces an out-of-plane deformation of the chromophore of rhodopsin. In our model, the deformation is distributed along the backbone in the region C9-C14, rather than being concentrated in the torsion of a specific bond. In our model with the Cl$^-$ counterion, the chirality of this torsion had the opposite sign than in the experiment [31]. In the new model with the CH$_3$COO$^-$ counterion we have the correct chirality and again the torsional angles in the region C9-C14 present significant deviations from planarity with a total out-of-plane deformation of approximately 30 degrees. Some of these torsional angles have been recently measured by NMR [12] and they agree well with the theoretical prediction. We can fairly conclude that our model describes quite accurately the details of the ground state structure of the chromophore.

Starting with the optimized structure of 11-*cis*-retinal PSB + Cl$^-$ (Z hereafter), we have applied an external torsional field which drives the system through the transition state (TS) to the all-*trans* product (E thereafter), provided the field is strong enough to overcome the ground-state isomerization energy barrier. As the dihedral angle C10-C11-C12-C13 increases, the HOMO-LUMO gap becomes smaller and a level crossing occurs when the transition state is reached. After the level crossing has been reached, the external field was set to zero, and the system evolved under the effect of the internal forces only. The system has been equilibrated for about 175 fs and subsequently cooled by a simulated annealing procedure in order to reach the minimum energy structure of the E product. The back-isomerization of the all-*trans* product E to the 11-*cis* reagent Z can be induced by applying again a torsional field. The transition state for the back-isomerization is denoted by TSb. The level crossing in TS and TSb is reached for two different values of the dihedral angle. Nevertheless, remarkably similar features have been observed in the bond alternation patterns. In TS and TSb the conjugation defect is displaced toward the center of the molecule, and located around C9. The displacement of the conjugation defect results in the inversion of the bond alternation in the tail of the molecule. The two transition states TS and TSb present different distributions of dihedral angles with essentially the same bond alternation patterns. Thus, the transition state is not characterized by a specific C10-C11-C12-C13 dihedral angle, which is usually assumed as the relevant reaction coordinate, but by a specific distribution of bond lengths, which reflects the displacement of the conjugation defect. The position of the defect is associated with the location of the excess positive charge along the conjugated chain. Prior to photoisomerization, it is located at the Schiff base end, upon excitation it is transferred to the head of the molecule, while in the transition state it is in the center of the molecule. We notice that the C11-C12 bond at the transition state is considerably elongated. This appears to be a necessary condition for the isomerization to occur.

In our *ab-initio* molecular dynamics simulations, the torsional field was set to zero after overcoming the energy barrier, i.e. after reaching the transition state. The subsequent evolution of the system toward the product E corresponds with a simulation of the chromophore relaxation after the non-radiative transition from the excited-state to the ground state potential. In Fig. 3 we report the time-dependent variation of the differences between adjacent C-C bond lengths calculated from the ionic trajectories during the CPMD simulation. As the system approaches the transition state TS, the conjugation defect is displaced from the Schiff base end toward the ring, causing a bond alternation inversion in the fashion of a propagating charged soliton. The torsional field was set to zero after 110 fs. When the molecule relaxes toward the all-*trans* product E, the conjugation defect coherently returns to the tail of the molecule. In the molecular dynamics simulation the coherent propagation of this defect is apparently associated with a propagation of the excess positive charge. We observe that the propagation of the

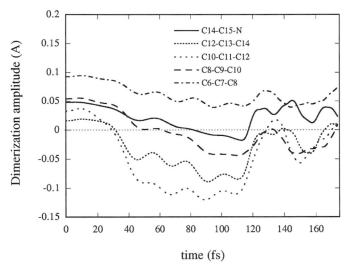

Fig. 3: Dimerization amplitude as a function of time for the direct 11-cis to all-trans isomerization induced by an external torsional field V_0=80 kcal/mol. Adapted from Ref. [16].

solitonic polaron is strongly coupled to a slow collective vibrational mode of the chromophore. Damping of the soliton is attributed to coupling with out-of-plane vibrational degrees of freedom due to non-planarity of the chromophore [14]. The present study shows that a soliton propagation is intimately connected with the chromophore relaxation toward the photoproduct. The product E presents a BLA defect in the terminal region of the backbone that is similar to the reagent Z and much more delocalized along the chain, in line with NMR experiments [32]. Moreover, the torsional angles in the all-*trans* product indicate a severe strain in the backbone. Thus, the retinylidene chromophore of bathorhodopsin is not in a fully relaxed all-*trans* form. Noticeably, the highly distorted all-*trans* structure is obtained with relatively small atomic displacements from the 11-*cis* configuration. This is consistent with the very short time of formation of bathorhodopsin upon *cis*-to-*trans* isomerization. Our model for the bathorhodopsin has been revalidated recently by circular dichroism calculations [31].

The computed energy stored in this model bathorhodopsin PSB including the Cl⁻ counterion is 22 kcal/mol. The energy stored in the system is not related to the charge separation between chromophore and counterion, but rather to the strain in the highly distorted all-*trans* configuration. However, the quantitative estimate of the energy storage is likely to depend on a further refinement of the counterion structure and distance from the chromophore. In addition, water molecules in the binding pocket and interactions with other protein residues may also play a role. Nevertheless, the mechanisms described above depend on the fundamental physical properties of the system and are generally applicable. It is therefore tempting to conclude that the classically coherent dynamics that transpire from our calculations will be encountered more often in the near future in the description of fast isomerization reactions in conjugated chains.

2. ^{13}C *Chemical Shifts* In this section we present the calculated ^{13}C chemical shifts for the 11-*cis*-retinal PSB models of the rhodopsin chromophore. In order to check the

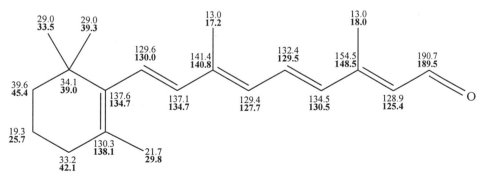

Fig. 4 : *Calculated (bold face) and experimental (regular)* ^{13}C *chemical shift values of all-trans-retinal.*

accuracy of the density functional, we have performed a test calculation for the all-*trans*-retinal compound. The results obtained with the B3LYP functional and a 6-31G basis set are presented in Fig. 4 and are compared with the experimental values [33]. The agreement between theory and experiment is very good with a root mean square (rms) error of 3.7 ppm. The correlation coefficient between the calculated and experimental chemical shifts is $\rho=0.981$. This result is consistent with a previous plane-wave pseudopotential calculation [18], thus showing that the computed chemical shifts are not very sensitive to the specific basis set used.

In the charged compounds, the presence and position of the counterion has a strong influence on the ^{13}C chemical shifts [18]. Indeed the experimental NMR spectra obtained in solution or for crystalline samples are clearly distinct [30]. In addition, the shifts in the spectra collected from crystalline samples depend on the counterion type [30]. In Fig. 5 we compare the experimental chemical shifts for rhodopsin [9] with the theoretical results obtained for the 11-*cis*-retinal PSB models with the Cl⁻ counterion and with the CH_3COO^- counterion. The agreement with the experiment improves significantly in the case of the acetate counterion that mimics the Glu$_{113}$ counterion in the NMR model and the crystalline structure [7] (see also Fig. 6): The rms error goes from 7.6 ppm in the case of the Cl⁻ counterion at 4 Å from C12, to 5.5 ppm in the case of the CH_3COO^- counterion. In addition, the correlation coefficient between the calculated and experimental chemical shift values improves from $\rho=0.880$ in the case of the Cl⁻ counterion to $\rho=0.972$ in the case of the CH_3COO^- counterion. The agreement improves particularly for the even numbered carbon sites and for the C5. It is known

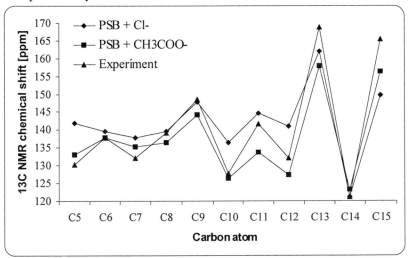

Fig. 5 : Comparison of the calculated ^{13}C chemical shifts in the conjugated chain of the 11-cis-retinal PSB + Cl⁻ and the 11-cis-retinal PSB + CH₃COO⁻ model compounds with the experimental values of bovine rhodopsin.

that the C5 chemical shift depends strongly on the torsional angle between the β-ionone ring and the conjugated chain [33]. Therefore the larger error on the C5 site for the model with Cl⁻ can be correlated with the underestimation of the C5-C6-C7-C8 torsional angle: In fact the dihedral angle C5-C6-C7-C8 goes from 37 degrees in the model with the chloride counterion to 55 degrees in the model with the acetate counterion. The large downfield chemical shifts measured at the C13 and C15 positions are not reproduced by either of our minimal models. Although the calculated chemical shift at the C15 improves in the model with the acetate counterion, the calculated chemical shift at the C13 is more in agreement with the experiment in the model with the chloride counterion. This indicates that further improvement in the counterion geometry and refinement of the X-ray structure may be necessary. It is likely that other charged residues close to the chromophore may contribute to the large downfield shift at the C13 as suggested in Ref. [22]. Recently, from a ^{15}N MAS NMR study, experimental evidence has been found for a complex counterion in which the carboxylate side chain of the glutamate residue interacts with the proton on the Schiff base nitrogen via a hydrogen-bonded bridge involving one or several water molecules [34]. Improved knowledge of the protein structure should allow us to extend our model in the future by including other charged residues and to check their relevance in explaining the measured chemical shifts.

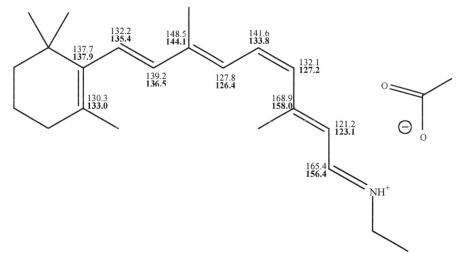

Fig. 6 : Calculated (bold face) and experimental (regular) ^{13}C chemical shift values for the carbon atoms in the conjugated chain of the retinylidene chromophore of rhodopsin.

Conclusions

We have used Density Functional Theory and Car-Parrinello molecular dynamics to study a retinylidene PSB model for the chromophore of rhodopsin. The equilibrium structure and the ^{13}C NMR chemical shifts of the model are in close quantitative agreement with the available experimental data.

We find that, as an effect of the positive charge delocalization, the amplitude of the bond length alternation between single and double carbon bonds along the chromophore is strongly reduced in the vicinity of the protonated Schiff base nitrogen. When we drive the chromophore along the reaction coordinate over the isomerization energy barrier, the excess positive charge is displaced from the Schiff base toward the ß-ionone ring. The transition state is characterized by a specific bond length pattern with the C11-C12 bond considerably elongated. Relaxation of the system from the transition state to the all-*trans* photoproduct (bathorhodopsin) induces a classically coherent propagation of the positively charged conjugation defect in the fashion of a charged soliton. The all-*trans* primary photoproduct is highly distorted and the energy is stored mainly in torsional strain.

The ^{13}C chemical shifts are sensitive to the nature and position of the counterion. The agreement with experiment improves when introducing in the model an acetate counterion positioned according to the recently determined crystalline structure. It is likely that other charged residues close to the chromophore may contribute to the large downfield shift at the C13 as suggested in Ref. [22]. An improved knowledge of the protein structure should allow us to extend this model in the future by including other charged residues and to check their relevance in explaining the measured chemical shifts.

Acknowledgements

We acknowledge F. Mauri and P. Giannozzi for sharing with us some unpublished chemical shifts results.

References:

[1]   Birge, R.R., Biochim. Biophys. Acta 1016 (1990) 293.
[2]   Mathies, R.A. and Lugtenburg, J. In Stavenga, D.G., de Grip, W.J., and Pugh Jr, E.N. (eds.) Handbook of Biological Physics Vol.3, Elsevier Science BV, 2000, p. 55
[3]   Schoenlein, R.W., Peteanu, L.A., Mathies, R.A. and Shank, C.V., Science 254 (1991) 412.
[4]   Birge, R.R., Einterz, C.M., Knapp, H.M. and Murray, L.P., Biophys. J. 53 (1988), 367
[5]   Shieh, T., Han, M., Sakmar, T.P. and Smith, S.O., J. Mol. Biol. 269 (1997) 373.
[6]   Schick, G.A., Cooper, T.M., Holloway, R.A., Murray, L.P. and Birge, R.R., Biochemistry 26 (1987) 2556.
[7]   Palczewski, K., Kumasaka, T., Hore, T., Behnke, C.A., Motoshima, H., Fox, B.A., Le Trong, I., Teller, D.C., Okada, T., Stenkamp, R.E., Yamamoto, M. and Miyano, M., Science 289 (2000) 739.
[8]   Palings, I., Pardoen, J.A., van den Berg, E., Winkel, C., Lugtenburg, J. and Mathies, R.A., Biochemistry 26 (1987) 2544.
[9]   Smith, S.O., Palings, I. Copié, V., Raleigh, D.P., Courtin, J. , Pardoen, J.A., Lugtenburg, J., Mathies, R.A. and Griffin, R.G., Biochemistry 26 (1987) 1606.

[10]   Smith, S.O., Palings, I., Miley, M.E., Courtin, J., de Groot, H.J.M., Lugtenburg, J.,
        Mathies, R.A. and Griffin, R.G., Biochemistry 29 (1990) 8158.
[11]   Verdegem, P.J.E., Bovee-Geurts, P.H.M., de Grip, W.J., Lugtenburg, J. and de Groot,
        H.J.M., Biochemistry 38 (1999) 11316.
[12]   Feng, X., Verdegem, P.J.E., Lee, Y.K., Sandström, D., Edén, M., Bovee-Geurts, P., de
        Grip, W.J., Lugtenburg, J., de Groot, H.J.M. and Levitt, M.H., J. Am. Chem. Soc. 119
        (1997) 6853.
[13]   Car, R. and Parrinello, M., Phys. Rev. Lett. 55 (1985) 2471.
[14]   Buda, F., de Groot, H.J.M. and Bifone, A., Phys. Rev. Lett. 77 (1996) 4474.
[15]   Bifone, A., de Groot, H.J.M. and Buda, F., J. Phys. Chem. B 101 (1997) 2954.
[16]   La Penna, G., Buda, F., Bifone, A. and de Groot, H.J.M., Chem. Phys. Lett. 294 (1998)
        447.
[17]   Sakmar, T.P., Franke, R.R. and Khorana, H.G., Proc. Natl. Acad. Sci. USA 86 (1989)
        8309
[18]   Buda, F., Giannozzi, P. and Mauri, F., J. Phys. Chem. B 104 (2000) 9048.
[19]   Becke, A.D., Phys. Rev. A 38 (1988) 3098
[20]   Perdew, J.P., Phys. Rev. B 33 (1986) 8822
[21]   Vanderbilt, D., Phys. Rev. B 41 (1990) 7892
[22]   Han, M. and Smith, S.O., Biochemistry 34 (1995) 1425
[23]   Mauri, F., Pfrommer, B.G. and Louie, S.G., Phys. Rev. Lett. 77 (1996) 5300.
[24]   Kleinman, L. and Bylander, D.M., Phys. Rev. Lett. 48 (1982) 1425
[25]   CPMD, J. Hutter, A. Alavi, T. Deutch, M. Bernasconi, St. Goedecker, D. Marx, M.
        Tuckerman, M. Parrinello, MPI für Festkörperforschung and IBM Zürich Research
        Laboratory 1995-1999
[26]   Gaussian 98, Revision A.5
        M.J. Frisch, G.W. Trucks, H.B. Schlegel, G.E. Scuseria, M.A. Robb, J.R. Cheeseman,
        V.G. Zakrzewski, J.A. Montgomery, Jr., R.E. Stratmann, J.C. Burant, S. Dapprich, J.M.
        Millam, A.D. Daniels, K.N. Kudin, M.C. Strain, O. Farkas, J. Tomasi, V. Barone, M.
        Cossi, R. Cammi, B. Mennucci, C. Pomelli, C. Adamo, S. Clifford, J. Ochterski, G.A.
        Petersson, P.Y. Ayala, Q. Cui, K. Morokuma, D.K. Malick, A.D. Rabuck, K.
        Raghavachari, J.B. Foresman, J. Cioslowski, J.V. Ortiz, B.B. Stefanov, G. Lui, A.
        Liashenko, P. Piskorz, I. Komaromi, R. Gomperts, R.L Martin, D.J. Fox, T. Keith, M.A.
        Al-Laham, C.Y. Peng, A. Nanayakkara, C. Gonzalez, M. Challacombe, P.M.W. Gill, B.
        Johnson, W. Chen, M.W. Wong, J.L. Andres, C. Gonzalez, M. Head-Gordon, E.S.
        Replogle, and J.A. Pople, Gaussian, Inc., Pittsburgh PA, 1998
[27]   Becke, A.D, J. Chem. Phys. 98 (1993) 5648
[28]   Lee, C., Yang, W. and Parr, R.G., Phys. Rev. B 37 (1988) 785
[29]   Ditchfield, R., Mol. Phys. 27 (1974) 789
[30]   Elia, G.R., Childs, R.F., Britten, J.F., Yang, D.S.C. and Santarsiero, B.D., Can. J. Chem.
        74 (1996) 591.
[31]   Buß, V., Kolster, K., Terstegen, F., and Vahrenhorst, R., Angew. Chem. Int. Ed. 37
        (1998) 1893
[32]   Smith, S.O., Courtin, J., de Groot, H., Gebhard, R. and Lugtenburg, J., Biochemistry 30
        (1991) 7409.
[33]   Harbison, G.S., Mulder, P.P.J., Pardoen, H., Lugtenburg, J., Herzfeld, J. and Griffin,
        R.G., J. Am. Chem. Soc. 107 (1985) 4809.

[34]   Creemers, A.F.L., Klaassen, C.H.W., Bovee-Geurts, P.H.M., Kelle, R., Kragl, U., Raap, J., de Grip, W.J., Lugtenburg, J., and de Groot, H.J.M., Biochemistry 38 (1999), 7195

Refolded G protein-coupled receptors from *E. coli* inclusion bodies

Hans Kiefer[a] , Klaus Maier[a] and Reiner Vogel[b]

[a]*m-phasys GmbH, Vor dem Kreuzberg 17, D-72070 Tübingen, Germany.*
[b]*Universität Freiburg, Institut für Biophysik, Albertstrasse 23, D-79104 Freiburg, Germany*

Introduction

G protein-coupled receptors (GPCRs) are the largest receptor family in eukaryotes with about 2000 representatives in the human genome. Of these, about 1500 GPCRs transmit signals in intracellular communication, while the remaining part are involved in the recognition of sensory signals such as odorants, gustatory substances and, through the visual pigment rhodopsin, light [1, 2].

It has been estimated that every other pharmaceutical drug acts on a GPCR, which makes these proteins a very attractive research object in the pharmaceutical industry. Despite enormous efforts, structural information has been limited until recently to a low-resolution map reconstructed from electron microscopy [3]. Thanks to Palczewski and colleagues [4] who could successfully grow well-diffracting rhodopsin crystals, an atomic-resolution structure has now become available. While the transmembrane portions of this structure may serve as a template in modelling other GPCRs, the loop regions are in general poorly conserved among different receptors. Moreover, GPCRs bind to an enormous variety of ligands ranging from small molecules to globular proteins. Therefore, the ligand binding sites are necessarily highly variable and many more experimental structures of different GPCRs complexed with ligands will be needed to understand the ligand-gated activation mechanism underlying GPCR function.

Why is so little structural information about GPCRs available? Their natural abundance in tissue is low (with rhodopsin being the notable exception) and expression in heterologous systems is often difficult in terms of yield or homogeneity. In addition, GPCRs, like membrane proteins in general, do not readily produce crystals diffracting to high resolution. Extensive fine-tuning of purification conditions and the choice of a

S.R.Kiihne and H.J.M.deGroot (eds.), Perspectives on Solid State NMR in Biology, 123-130.

detergent suitable for crystallization may require many years of experimental biochemistry.

Moreover, the crystallization of integral membrane proteins derived from the cytoplasmic membrane of eukaryotic cells seems to be an especially difficult task. Indeed, rhodopsin is the only protein from this source for which a structure has been solved. All other membrane protein structures belong to bacterial or organelle membrane proteins. Is this observation related only to the low abundance of cytoplasmic membrane proteins? Probably not, since heterologous expression systems such as insect cells or yeast sometimes yield appreciable amounts of GPCRs as well as other membrane proteins. A clue to what could be the additional problem comes from a mass spectroscopical analysis of membarne proteins by Whitelegge et al. [5]. They found that rhodopsin from retinal disk membranes is a mixture of about 50 different species which are probably glycosylated, palmitoylated and phosphorylated to various extents. In striking contrast, proteins from bacterial or chloroplast membranes produced a single peak in MS.

In this paper, we present an approach leading to the production of large amounts of purified receptor protein from a bacterial expression system. We chose *E. coli* as a host for expression of a fusion protein containing a receptor sequence as well as an N-terminal glutathione-S-transferase (GST) moiety and a C-terminal 6-His tag, thus allowing rapid purification. Expression of these constructs results in the formation of inclusion bodies containing misfolded protein. Although this fact is a major drawback which implies that special techniques must be developed to refold the protein into its native state, inclusion bodies also have advantages: The yield of expressed protein reaches up to ten percent of the cellular protein which is at least 100 fold higher than in any other GPCR expression system. Inclusion body protein is insoluble in neutral detergents which allows removal of soluble and membrane proteins by centrifugation and thus facilitates protein purification [6].

Although this expression system was initially developed to produce protein suitable for X-ray crystallography, it may also be advantageous in NMR-based structural investigations. Homogeneity of the sample is required to obtaining spectra of high quality. In addition, E. coli is certainly the preferred expression system for NMR studies, since isotope labelling is much easier and cheaper than in any other expression system as long as high expression levels can be achieved.

Two topics are addressed in the following. (i) We describe how to obtain high expression levels of GPCRs leading to inclusion bodies. (ii) We outline how to refold GPCRs from inclusion bodies into their native states, and we show evidence for ligand binding activity of refolded GPCRs.

Methods

Expression. All GPCRs were cloned into the same pGEX-His fusion vector (Fig. 1). The plasmids were transformed into *E. coli* BL21. Expression levels were determined by purifying corresponding amounts of fusion proteins from small cultures (1 ml)

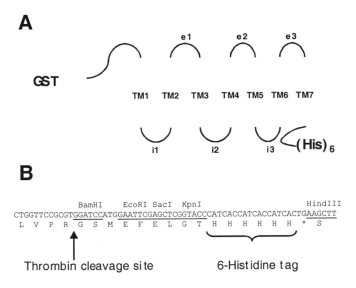

A

GST

e1   e2   e3

TM1   TM2   TM3   TM4 TM5 TM6 TM7

i1   i2   i3   (His)₆

B

```
          BamHI      EcoRI SacI  KpnI                              HindIII
CTGGTTCCGCGTGGATCCATGGAATTCGAGCTCGGTACCCATCACCATCACCATCACTGAAGCTT
  L   V   P   R   G   S   M   E   F   E   L   G   T   H   H   H   H   H   H   *   S
```

          ↑
Thrombin cleavage site   6-Histidine tag

*Fig. 1. (A) Scheme of GST-GPCR fusion constructs. (B) Sequence of thrombin cleavage site,
multiple cloning site, and (His)₆ tag.*

induced for 4 hrs with 0.1 mM IPTG and quantifying the protein content by 2D
densitometry of an SDS-PAGE stained with Coomassie Blue.

Multiple linear regression analysis. Λ statistical analysis was performed to detect a
correlation between the occurrence of positively charged residues and expression levels.
For every receptor, a dataset with six variables corresponding to the positive charge
content in the six loop regions (3 intracellular and 3 extracellular loops) was calculated.
The variables, denoted "weighted fraction of positive charges (WFPC)", take only the
variation between 0% and 25% of positive charge content in every loop into account
and are defined as WFPC = 1 for FPC > 0.25, and WFPC = 4 x (FPC) for FPC ≤ 0.25.
FPC, the fraction of positive charges, is simply the number of Arg + Lys divided by the
total number of residues in each loop. The WFPCs were chosen as independent
variables for multiple linear regression analysis (MLR) using the program Statistica
with the expression level as dependent variable.

Refolding.   The procedure used to refold GPCRs contains three basic steps:
Solubilization of the inclusion body protein, transfer into a neutral detergent, thrombin
cleavage, purification on a Ni-NTA column. For solubilization, N-lauroyl sarcosine
(sarcosyl) is added to the resuspended inclusion bodies. The concentration required to
solubilize the protein has to be determined and is usually between 1 and 3%. Next, a
second, milder detergent is added to the dissolved protein and refolding is induced.
Choice of the most suitable detergent is important at this step and varies among

different receptors. It can only be found by screening a large set of detergents since no rationale able to predict the behaviour of a given receptor in different detergents is available up to now. Sometimes, detergents used in the literature to stabilize the corresponding GPCR when expressed in its native state may provide useful starting conditions. Digitonin has proven to be especially successful at inducing refolding in several cases. After adding the mild detergent, thrombin is added and the solution is kept at 20°C until the fusion protein is completely digested (5 min to 20 hrs). The cleaved protein is purifed on a Ni-NTA column and eluted with imidazole.

Results and Discussion

We found that expression levels of different GST/receptor fusions varied between 0.1 and 10% of the total cellular protein, depending on the specific receptor sequence. Low expression levels were correlated with a toxic effect of protein expression; Strains that expressed poorly also stopped growing immediately after induction with IPTG, while strains yielding high levels of fusion protein continued to grow.

To understand this phenomenon, we looked for specific features in the protein sequence which could cause the toxic effect. Heterologous expression of membrane proteins in *E. coli* is often toxic, and it has been observed that toxicity is not related to either transcription or translation, but occurs at a later stage such as membrane insertion.

Membrane proteins insert into the membrane according to the "positive inside rule" which was discovered by statistical analysis of protein sequences and subsequently confirmed by site-directed mutagenesis studies [7]. This rule predicts that loops connecting transmembrane segments stay on the cytoplasmic side of the membrane if they contain at least 20% positively charged residues (Arg and Lys). Loops with less positive charge content can be translocated across the membrane. This rule is valid both in pro- and eukaryotes, but there are some important differences. Namely, in bacteria, it strictly applies to nearly every single loop, while in eukaryotes, many exceptions occur, and the charge bias is visible only when many sequences are averaged [8]. Therefore, inefficient insertion into the membrane could be the reason for toxicity.

We measured expression levels of 15 different GPCR sequences and attempted to relate the data to the positive charge content in the loop regions. A significant correlation was observed if only the variation between 0 and 25% of positive charge was taken into account. This was expected from the positive inside rule, as loops with a charge content of 25% or higher always remain intracellular. In other words, the variation above 25% is not determining membrane insertion. The correlation was established by multiple linear regression analysis, and the regression showed a positive correlation with the expression level in four out of six cases, supporting the hypothesis that loops with a low charge content are toxic [9].

The data suggest that expression levels of GST-GPCR fusions can be improved by increasing the positive charge content in loops containing no or few charges. Site-directed mutagenesis experiments have confirmed this hypothesis in several cases (Fig. 2).

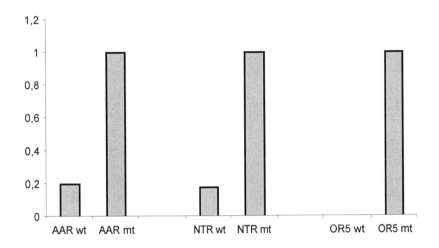

Fig. 2. Effect of introducing positively charged residues into loop regions. The relative expression levels of three receptor wild-type ("wt") sequences and the corresponding mutants ("mt") with additional positively charged residues in one loop are shown. AAR, alpha1B adrenergic receptor (hamster); OR5, odorant receptor 5 (rat); NTR, neurotensin receptor (rat).

The procedure used to refold GPCRs is related to established techniques commonly used to refold soluble proteins from inclusion bodies. For these proteins, inclusion bodies are normally solubilized by a chaotropic agent such as urea or guanidinium chloride, yielding completely unfolded polypeptide chains. Subsequently, the protein is diluted into a suitable "folding buffer", where it adopts the native conformation. Obviously, the composition of the folding buffer is crucial for the whole procedure. The major side-reaction which reduces refolding yields is irreversible aggregation. To kinetically favour folding over aggregation, one can work at dilute protein concentrations, since aggregation in contrast to folding is a second-order reaction. However, the concentration required to obtain appreciable yields may be so low that subsequent concentration of the protein is impractical. Therefore, much effort has been put into additives in the folding buffer which reduce aggregation. Examples of such additives are arginine, detergents and polyethylene glycol [10].

Membrane proteins are not soluble without detergent. Therefore, chaotropic agents are not able to solubilize inclusion bodies of membrane proteins unless a detergent is added simultaneously. On the other hand, harsh detergents with a net charge can solubilize such inclusion bodies without the addition of urea. However, a harsh detergent is denaturing and has to be replaced by a mild detergent which allows the protein to adopt its native conformation. Since detergents form micelles and strongly bind to membrane proteins, complete removal of the first detergent and replacement by the second one is only possible when the protein is bound to a column.

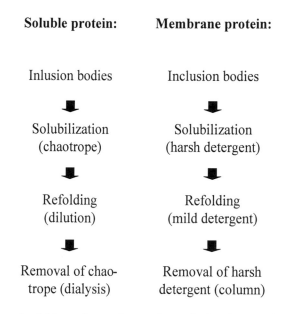

Soluble protein:	Membrane protein:

Fig. 3. *Comparison of refolding schemes from inclusion bodies for soluble and membrane proteins.*

The routes towards folded soluble and membrane proteins are shown schematically in Fig. 3.

Digitonin is widely used to stabilize GPCRs in their native conformation and has been used by us to refold several GPCRs. The drawbacks of digitonin are that it is a natural product composed of two isomers, it only becomes water-soluble upon boiling, and its properties depend on the supplier and the batch. Therefore, digitonin will have to be replaced by other detergents in many applications.

The first evidence for activity of a refolded GPCR was obtained with the odorant receptor OR5 [11]. Although the natural ligand lilial is hydrophobic and therefore cannot be used in standard ligand binding assays, a derived photoaffinity label (BDCA) reacted covalently with OR5 upon illumination. This reaction was inhibited in a concentration-dependent manner by lilial, indicating that both molecules bind to the same site on the receptor.

Subsequently, a shark A_0-adenosine receptor was purified and refolded as described above and adenosine binding could be measured by a centrifugation method. Receptor, reconstituted in lipid vesicles, was incubated with ^{3}H-adenosine, and bound radioactivity was separated from free adenosine by ultracentrifugation (Fig. 4). The results allow quantification of the yield of refolded protein which at present accounts for ca. 1% of the total receptor protein.

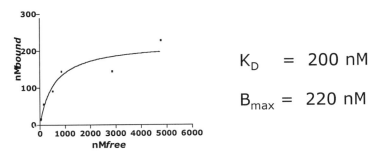

$$K_D = 200 \text{ nM}$$

$$B_{max} = 220 \text{ nM}$$

Fig. 4. Ligand binding assay of reconstituted shark A0 adenosine receptor. The proteoliposomes were incubated with radioactive adenosine and separated from unbound ligand by centrifugation. The K_D ans B_{max} were determined by non-linear curve-fitting.

If binding of the solubilized receptor to a ligand affinity column is measured, about 30% of the protein binds and can be eluted by competition with adenosine. This implies that refolding takes place already during the solubilization and purification procedure and not during reconstitution into a lipid membrane. For reasons we ignore, most of the refolded protein present in the solubilized state is lost during reconstitution. However, as we want to obtain a fraction highly enriched in active receptor and suitable for biophysical studies and crystallization screenings, we prefer to work in the detergent-solubilized state.

Conclusions

While refolding of soluble proteins from inclusion bodies has become a standard laboratory technique, this is not yet true for membrane proteins. The latter are in general considered to be more difficult to handle and less stable than soluble proteins. Loss of activity is usually a consequence of aggregation which is considered "irreversible". This notion has probably discouraged some researchers from even attempting to fold membrane proteins from a non-native state. However, examples of successfully refolded membrane proteins have been in the literature for many years [12-14]. Detergents can perform a double task since they bind readily to the hydrophobic segments of membrane proteins, thereby inducing the formation of alpha helices and at the same time preventing aggregation. Indeed, the far-UV CD spectrum of a GPCR in SDS is very similar to that of the same protein dissolved in a mild detergent, implying that the secondary structure is formed although the detergent is considered to be denaturing. By replacing the harsh with a mild detergent, the protein probably only has to form the right interhelical contacts present in the native structure. This is accomplished when the detergent micelle is both flexible enough to allow rapid conformational changes and stable enough to prevent hydrophobic contacts between proteins in different micelles. Unfortunately, these two effects are in principle contradictory: detergents with a low CMC form stable micelles at the cost of flexibility and *vice versa*. This is probably the reason why choosing the right detergent is so critical, and why there is no general recipe for every GPCR or membrane protein.

The main advantage of producing GPCRs in *E. coli* inclusion bodies is the large amount and the homogeneity of the purified protein. Although the refolding procedure may seem tedious, we believe that this route is a promising approach for all kind of applications where large amounts of purified protein are needed. However, the ultimate proof of our concept will have to be the first structure of a refolded membrane protein solved either by X-ray crystallography or by NMR.

References

[1]   Bourne, H.R., Curr Opin Cell Biol 9 (1997) 134.
[2]   Gether, U. and Kobilka, B.K., J Biol Chem 273 (1998) 17979.
[3]   Schertler, G.F. and Hargrave, P.A., Proc Natl Acad Sci U S A 92 (1995) 11578.
[4]   Palczewski, K., et al., Science 289 (2000) 739.
[5]   Whitelegge, J.P., Gundersen, C.B., and Faull, K.F., Protein Sci 7 (1998) 1423.
[6]   Rudolph, R. and Lilie, H., Faseb J 10 (1996) 49.
[7]   von Heijne, G., Embo J 5 (1986) 3021.
[8]   Wallin, E. and von Heijne, G., Protein Eng 8 (1995) 693.
[9]   Kiefer, H., Vogel, R., and Maier, K., Receptors Channels 7 (2000) 109.
[10]   Lilie, H., Schwarz, E., and Rudolph, R., Curr Opin Biotechnol 9 (1998) 497.
[11]   Kiefer, H., et al., Biochemistry 35 (1996) 16077.
[12]   Huang, K.S., et al., J Biol Chem 256 (1981) 3802.
[13]   Fiermonte, G., Walker, J.E., and Palmieri, F., Biochem J 294 (1993) 293.
[14]   Rogl, H., et al., FEBS Lett 432 (1998) 21.

Semliki Forest virus vectors: versatile tools for efficient large-scale expression of membrane receptors

Kenneth Lundstrom

F. Hoffmann-La Roche, CNS Department, Bldg 69/440, CH-4070 Basel, Switzerland

Introduction

One of the prerequisites for success in structural biology is to have access to large quantities of the protein of interest to enable purification and crystallization and furthermore to carry out high-resolution structural analysis. It is well known that only rarely, as in the case of rhodopsin, the protein concentration in native tissue is high enough to allow direct purification of material for structural studies. Particularly, transmembrane receptors are expressed at a rather low density and so far, among seven transmembrane receptors, only bacteriorhodopsin [1] and bovine rhodopsin [2, 3] have been successfully purified, crystallized and high-resolution structures resolved. There is, however, an enormous interest in the structural biology of receptors belonging to the family of G-protein coupled receptors (GPCRs) because many of these are involved in neurotransmission and important signal transduction events specifically in nerve cells. For this reason, GPCRs are the target molecules for intensive research to develop novel drugs for many central nervous system disorders including anxiety, depression and memory dysfunction.

In attempts to achieve efficient expression of recombinant GPCRs many different host organisms have been employed. Prokaryotic expression vectors using *E. coli* as host cells have been applied for the expression of several GPCRs [4]. However, successful expression has generally required modification of the receptor gene by either deletion of some or all transmembrane regions, or expression of the GPCR as a fusion to a bacterial protein to stabilize and protect the structure against bacterial proteases. The lack of post-translational modifications required by mammalian GPCRs is a major drawback with *E. coli*-based expression. Yeast expression systems have turned out to be relatively efficient for GPCR expression, although the glycosylation pattern is quite different from mammalian cells and can therefore be a disadvantage for expression of mammalian receptors [5]. Baculovirus vectors have allowed GPCR expression in insect

S.R.Kühne and H.J.M.deGroot (eds.), Perspectives on Solid State NMR in Biology, 131-139.

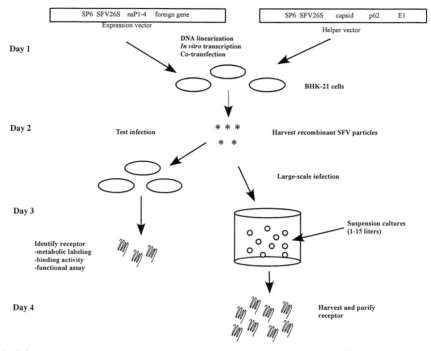

Fig. 1. Schematic presentation of the SFV expression system. Recombinant SFV particles are generated in BHK-221 cells in <2 days. Replication-deficient SFV vectors are used for small and large-scale expression of recombinant proteins.

cells to relatively high levels (B_{max} values of 10-40 pmol receptor / mg protein) [6]. The drawbacks have mainly been the relatively time-consuming virus production and the restricted expression in insect cells. Quite a few attempts have been made to overexpress GPCRs from transient or stable mammalian expression vectors [7]. Although successful receptor expression has been obtained the methods are rather time-consuming, labor intensive and inefficient.

There has therefore been an urgent need for the development of more efficient mammalian expression systems GPCRs. Viral vectors have generally showed a great potential for recombinant gene expression in mammalian host cells due to their efficient gene delivery capacity and strong promoter activity. Especially, an expression system based on the Semliki Forest virus (SFV) genome has demonstrated rapid high-level gene expression of topologically different proteins in a variety of mammalian cell lines [8]. The SFV system is based on two plasmid vectors: 1. An expression vector containing the nonstructural SFV genes (responsible for RNA replication) and the foreign gene(s) of interest inserted downstream of the strong 26S subgenomic promoter. 2. A helper vector with the structural SFV genes (capsid and membrane protein genes) required for the assembly of virions. High-titer (10^9-10^{10} infectious particles / ml) replication-deficient SFV particles are generated in less than two days by *in vitro* transcription of RNA from the expression and helper vectors followed by co-transfection into BHK cells (Fig. 1). *In vivo* packaged recombinant SFV particles are harvested 24 h later. The SFV system has turned out to be extremely efficient for expression of GPCRs [9]. The rapid generation of recombinant SFV has allowed large-scale production of recombinant receptors within 4 days.

Here is reviewed the use of SFV vectors for expression of several GPCRs. Studies on binding activity and functional coupling to G proteins are described. Emphasis is put on the large-scale expression of GPCRs and ligand-gated ion channels which has resulted in production of milligram quantities of receptors per liter culture. This has substantially facilitated receptor purification for structure-function studies.

Methods

Cell cultures. BHK-21, CHO-K1 and HEK293 cells were cultured in a 1:1 mixture of Dulbecco's modified F-12 medium (Gibco BRL) and Iscove's modified Dulbecco's medium (Gibco BRL) supplemented with glutamine and 10% fetal calf serum. CHO-K1 cells were adapted to growth in serum-free suspension culture in DHI-medium (Gibco BRL) as earlier described [10].

Generation of recombinant SFV particles. SFV vectors containing various GPCRs, the mouse serotonin 5-HT$_3$ receptor and the pSFV-Helper2 [11] vector were linearized with either *Spe*I (pSFV1 based clones) or *Nru*I (pSFV2gen based clones) restriction endonucleases. *In vitro* transcribed RNA from expression and helper vectors were co-electroporated into BHK-21 cells and *in vivo* packaged SFV particles were harvested 24h later as described earlier [9]. Virus stocks were filter-sterilized, activated by α-chymotrypsin (Roche Molecular Biochemicals) treatment and the reaction terminated by addition of the trypsin inhibitor aprotinin (Sigma) prior to use. The α-chymotrypsin activation is carried out because virus generated from pSFV-helper2 vector will result in conditionally infectious particles due to three point mutations located in the p62 precursor for E2 and E3 membrane proteins [11]. Approximate virus titers were estimated from infection of known numbers of BHK-21 cells with serial dilutions of virus stocks. Infected cells were recognized by reporter gene expression (β-galactosidase, GFP) or the typical change in morphology caused by SFV.

SFV-mediated expression of GPCRs and ligand-gated ion channels. BHK cells were infected with SFV-GPCR and SFV-5-HT$_3$ vectors and metabolic labeling with [^{35}S]-methionine was carried out at 16 h post-infection. The recombinant receptor expression was verified by SDS-PAGE and visualized by autoradiography. Saturation binding experiments were performed on membrane preparates or whole-cell samples of SFV-infected BHK-21, CHO-K1 and HEK293 cells [9]. Functional coupling to G-proteins was assayed by measurement of intracellular Ca^{2+}-release [9], inositol phosphate accumulation [12], cAMP stimulation [13] and radioactive GTPγS binding [14].

Large-scale SFV expression and protein purification. BHK-21 and CHO-K1 cells were adapted to growth in suspension cultures [15]. Large-scale SFV-infections were performed in bioreactors with volumes up to 11.5 liters [16]. Generally, relatively high multiplicity of infection (MOI), 10-30, was used to achieve maximal expression levels. Although an MOI of 5 was sufficient to infect all cells in culture, higher expression levels were obtained with increased MOI values, most likely due to multiple infections. To obtain maximal expression levels, cells in the exponential phase were infected at a density of 7×10^5 cells / ml. Recently, BHK-21, CHO-K1 and HEK293 cells were adapted to growth in serum-free suspension cultures, which has significantly reduced the costs of recombinant protein production [10]. The serum-free medium had no effect on expression levels. Spinner flasks have been used for rapid receptor production in the range of 1-6 liter culture volumes.

Results and Discussion

SFV vectors have been successfully used for expression of GPCRs and ligand-gated ion channels in a variety of mammalian host cells [9, 17]. Many common mammalian cell lines (BHK-21, CHO-K1, COS7, HEK293, etc.) as well as primary cell cultures (fibroblasts, hepatocytes, neurons) are susceptible to SFV-infection. At the present time, more than fifty different GPCRs have been successfully expressed from SFV vectors (Table 1). Highly specific ligand binding has been obtained for many GPCRs. The majority of GPCRs expressed have shown B_{max} values in the range of 50 pmoles receptor / mg membrane protein, which is approximately 20-50 fold higher than obtained for any other mammalian expression system. Some receptors like the human adenosine A3 receptor expressed at modest levels (B_{max} 1 pmol/mg), but this receptor has been particularly difficult to express using conventional techniques [18]. Some receptors like the human neurokinin-1, the human α_{2C}-adrenergic and the rat histamine H2 receptors have been expressed at extreme levels (up to 200 pmol/mg). Saturation binding experiments performed on whole cells indicated that receptor densities of more than 6×10^6 receptors / cell were achieved [9]. One reason for the high expression level obtained from SFV vectors is the efficient amplification of the viral RNA molecules (including the foreign gene of interest) directly in the cytoplasm of mammalian host cells, which will result in an estimated 200,000 copies of RNA per cell. Because SFV has a positive(+)-strand RNA genome, the amplified RNA copies can directly act as mRNA molecules for efficient translation of recombinant proteins. Another feature of SFV is the shut down of host cell protein synthesis, which means that the majority of the synthesized proteins are of viral origin, and in this case the recombinant receptor (Fig. 2).

Table 1. SFV-mediated expression of GPCRs.

Receptor	B_{max} (pmol/mg)	Receptors/cell	Function	Reference
Adenosine A1, A2a, A2b	20-50	$1.0 - 3.0 \times 10^6$	cAMP	[21]
Adenosine A3	1.0	3.0×10^5	n.d.	[18]
Adrenergic α_{1B}	25	2.0×10^6	IP$_3$	[12]
Adrenergic α_{2C}	100-200	$5.0 - 10 \times 10^6$	n.d.	[21]
Adrenergic β_2	50-60	3.0×10^6	Ca^{2+}	[21]
Chemokine CCK-1, CCK-2	n.d.	$2.0 - 3.0 \times 10^5$	n.d	[21]
Dopamine D2, D4	n.d.	n.d.	n.d.	[21]
Dopamine D3	21	2×10^6	n.d.	[17]
Endothelin A, B	n.d.	n.d.	ERK	[21]
Galanin 1, 2, 3	20-50	$1.0 - 3.0 \times 10^6$	n.d.	[21]
Histamine H1, H2	50-100	$3.0 - 6.0 \times 10^6$	cAMP	[13]
Metabotropic Glutamate R1	120	6.0×10^6	n.d.	[21]
Metabotropic Glutamate R2,3	2-3	n.d.	cAMP, IP$_3$, EP	[22]
Metabotropic Glutamate R4,8	9-14	n.d.	GTPγS	[23]
Neuropeptide Y1, Y2	n.d.	n.d.	n.d.	[21]
Neurokinin-1, 2, 3	50-100	$3.0 - 8.0 \times 10^6$	Ca^{2+}	[9]
Olfactory I7, OR5, odr-10	n.d.	n.d.	n.d.	[24]
Opioid δ, κ, μ	20-50	$1.0 - 3.0 \times 106$	cAMP	[21]
Opsin Rh1	n.d.	n.d.	VP	[21]
Prostaglandin E2ep4	3.0	n.d.	GTPγS	[14]
Serotonin 5-HT1B, 1D, 6, 7	10-50	$1.0 - 3.0 \times 10^6$	cAMP	[21]

cAMP = cyclic AMP stimulation; Ca^{2+} = intracellular Ca^{2+}-release (Fura-2 or FLIPR); EP = Electrophysiology; ERK = extracellular signal-regulated kinase; GTPγS = GTPγS binding ; IP3 = inositol phosphate; VP = visual pigment formation

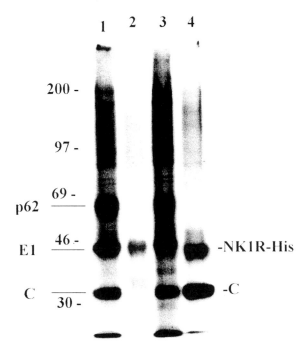

Fig. 2. Metabolic labeling of BHK-21 cells electroporated and infected with SFV vectors. BHK-21 cells electroporated with in vitro transcribed RNA from pSFV1-NK1R and pSFV-helper2 (lane 1) and pSFVCAP-NK1R-His and pSFV-helper2 (lane 3) were labeled with [35S]-methionine at 4 h post-electroporation. BHK-21 cells infected with recombinant SFV-NK1R (lane 2) and SFVCAP-NK1R-His (lane 4) were metabolically labeled at 16 h post-infection. C, SFV capsid; E1, SFV E1 membrane protein; p62, precursor for SFV E2 and E3 membrane proteins; NK1R-His, human neurokinin-1 receptor with C-terminal hexa-histidine tag. Molecular weight marker (kD) on the left.

In many cases, functional coupling of G-proteins has been demonstrated for GPCRs expressed from SFV vectors (Table 1). In these experiments, BHK-21, CHO-K1, COS7, HEK293, and even primary rat hippocampal neurons that were infected with SFV vectors carrying GPCRs have been subjected to assays for intracellular Ca^{2+}-release [9], inositol phosphate accumulation [12], cAMP stimulation [13] or GTPγS binding [14]. Generally, the maximal functional response was detected at 8-10 h post-infection followed by a decrease with time. This is due to the cell death induced by the SFV infection, but also because of shut down of endogenous gene expression, including G protein subunits. In combination with the extreme overexpression of recombinant GPCRs this will result in altered ratio of receptors vs. G-proteins leading to a decrease in functional activity. The lack of endogenous G proteins can be compensated by co-infection with SFV vectors expressing G protein subunits. By this procedure, increased functional responses were observed after quadruple infections of COS-7 cells with SFV vectors carrying the hamster α_{1B}-adrenergic receptor, $G_{\alpha q}$, $G_{\beta 2}$, and $G_{\gamma 2}$ subunits [12].

SFV-mediated expression of a ligand-gated ion channel, the mouse serotonin 5-HT$_3$ receptor, resulted in high specific binding activity with receptor densities of more than 3×10^6 receptors / cell. The pharmacological profile was identical to what had been observed for the 5-HT$_3$ receptor in native tissue. Moreover, functional activity could be demonstrated by whole-cell patch-clamp electrophysiological recordings resulting in high amplitude responses [16].

Because of the rather labor-intensive and time consuming procedure of infecting adherent cell cultures with SFV vectors and with the aim of large-scale receptor production, the SFV system was applied to suspension cultures of mammalian cells. BHK-21 and CHO-K1 cells adapted to growth in spinner flasks and bioreactors could be efficiently infected and demonstrated similar expression levels of recombinant proteins as in adherent cell cultures [15]. To facilitate the purification of the mouse serotonin 5-HT$_3$ receptor a hexa-histidine tag was introduced at the C-terminus. BHK-21 cells were infected with SFV-5-HT$_3$His virus at a multiplicity of infection (MOI) of 30 in an 11.5 liter bioreactor culture. Cells were harvested 16 h post-infection and analyzed for receptor expression and pharmacological properties. A B$_{max}$ value of 52 pmol/mg and a total receptor yield of 15 mg were obtained. Repeated bioreactor runs showed high reproducibility and yields of 1-2 mg receptor / liter culture were routinely achieved. The 5-HT$_3$ receptor was solubilized and efficiently purified via the histidine-tag. SDS-PAGE analysis of the subunit revealed a single 65 kD band suggesting a proper glycosylation of the receptor [19]. Size exclusion chromatography indicated that the size of the functional receptor complex was 280 kD, which is in good agreement with the postulated homopentameric structure for the 5-HT$_3$ receptor. Additionally, preliminary cryo-EM studies have confirmed the pentameric structure for the 5-HT$_3$ receptor.

To further increase the expression levels of the receptor gene of interest a fusion to the SFV capsid gene was engineered, because it has been demonstrated that a translation enhancement signal is located at the 5'end of the capsid gene [20]. Furthermore, fusions to full-length capsid result in autocatalytic cleavage of the capsid protein from the recombinant protein. The human neurokinin-1 receptor (hNK1R) gene was therefore fused to the capsid gene and expressed from the SFV vector resulting in 5-10 fold higher expression levels (5-10 mg receptor / liter) compared to the hNK1R expressed alone (Fig. 2). This fusion-GPCR protein has now been produced in large quantities (> 200 liters of CHO cultures). Cell fractionation indicated that only a minor number of the receptors were transported to the plasma membrane. However, the majority of the receptors residing in intracellular structures demonstrated functional binding activity (Lundstrom, unpublished results). This intracellular receptor localization could be confirmed by preliminary immuno-EM studies (Bohrmann and Lundstrom, unpublished results). The purified hNK1R is now subjected to 2D- and 3D-crystallography.

Conclusions

Semliki Forest virus (SFV) vectors have been developed for highly efficient transgene expression. Replication-deficient SFV particles can be rapidly generated at high titers. The broad host range of SFV has allowed recombinant protein expression in a large number of mammalian cell lines and primary cell cultures. SFV technology established for mammalian cells adapted to growth in suspension cultures has greatly facilitated large-scale recombinant receptor expression. Yields of several milligrams of receptor per liter culture have made purification of both the serotonin 5-HT$_3$ receptor and the neurokinin-1 receptor feasible. Based on preliminary studies on several other GPCRs (adrenergic, histamine and metabotropic glutamate receptors) similar high level expression was obtained. The purified receptors showed high functional binding activity. Preliminary structural characterization of the mouse serotonin 5-HT$_3$ receptor and the human neurokinin-1 receptor was possible. More detailed structural studies including cryo-EM, 2D- and 3D-crystallography are now in progress.

Acknowledgements

Dr. Horst Blasey (Glaxo Institute for Molecular Biology, Geneva) is acknowledged for his contribution to the large-scale SFV technology. I am thankful to Dr. Ernst-Jürgen Schlaeger (Roche, Basel) for his help with adaptation of mammalian cells to serum-free suspension cultures. Drs. Ruud Hovius and Horst Vogel (Swiss Federal Institute of Technology, Lausanne) are acknowledged for the purification and characterization of the mouse serotonin 5-HT$_3$ receptor. I am grateful to Dr. Raymond Stevens (Scripps, La Jolla) for his contribution to the purification of the human neurokinin-1 receptor. Ms. Raquel Herrador, Ms. Nicole Gnauck and Mr. Christophe Schweitzer (Roche, Basel) are acknowledged for excellent technical assistance with the virus preparation and large-scale expression work.

References

[1]   Schertler, G.F., Villa, C. and Henderson, R., Nature 362 (1993) 770.
[2]   Henderson, R., Baldwin, J.M., Ceska, T.A., Zemlin, F., Beckmann, E. and Downing, K.H. J., Mol. Biol. 213(1990), 899.
[3]   Palczewski, K., Kumasaka, T., Hori, T., Behnke, C.A., Motoshima, H., Fox, B.A., Le Tong, I., Teller, D.C., Okada, T., Stenkamp, R.E., Yamamoto, M. and Miyano, M. Science 289 (2000) 739.
[4]   Grisshammer, R. and Tate, C.G., Q. Rev. Biophys. 28 (1995) 315.
[5]   Weiss, H.M., Haase, W., Michel, H. and Reiländer, H. Biochem. J. 330 (1995) 1137.
[6]   Van den Broeck, J., Poels, J., Simonet, G., Dickens, L. and De Loof, A. Ann. N.Y. Acad. Sci. 839 (1998) 123.
[7]   Fraser, C.M., Arakawa, S., McCombie, W.R. and Venter, J.C. J. Biol. Chem. 264 (1989) 11754.
[8]   Liljeström, P. and Garoff, H. Bio/Technology 9 (1991) 1356.
[9]   Lundstrom, K., Mills, A., Buell, G., Allet, E., Adami, N. and Liljeström, P. Eur. J.

Biochem 224 (1994) 917.
[10] Schlaeger, E.-J. and Lundstrom, K. Cytotechnology 28 (1998) 205.
[11] Berglund, P., Sjöberg, M., Garoff, H., Atkins, G.J., Sheahan, B.J. and Liljeström, P. Bio/Technology 11 (1993) 916.
[12] Scheer, A., Björklöf K., Cotecchia, S. and Lundstrom, K. J. Receptor & Sign. Transd. Res. 19 (1999) 369.
[13] Hoffmann, M. Verzijl, D., Lundstrom, K., Simmen, U., Alewijnse, E.A., Timmerman, H. and Leurs, R. Submitted to Br. J. Pharmacol (2001).
[14] Marshall, F.H., Patel, K., Lundstrom, K., Camacho, J., Foord, S.M. and Lee, M.G. Br. J. Pharmacol. 121 (1997) 1673.
[15] Blasey, H.D., Lundstrom, K., Tate, S. and Bernard, A.R. Cytotechnology 24 (1997) 65.
[16] Lundstrom, K., Michel, A., Blasey, H., Bernard, A.R., Hovius, R., Vogel, H. and Surprenant, A. J. Receptor & Sign. Transd. Res. 17 (1997) 115.
[17] Lundstrom, K. J. Receptor & Sign. Transd. Res. 19 (1999) 673.
[18] Patel, M., Harris, C. and Lundstrom, K. Drug. Dev. Res. 40 (1997) 35.
[19] Hovius, R., Tairi, A.-P., Blasey, H., Bernard, A., Lundstrom, K. and Vogel, H. J. Neurochem. 70 (1998) 824.
[20] Sjöberg, M., Suomalainen, M. and Garoff, H. Bio/Technology 12 (1994) 1127.
[21] Lundstrom, unpublished results.
[22] Schweitzer, C., Kratzeisen, C., Dam, G., Lundstrom, K., Malherbe, P., Ohresser, S., Stadler, H., Wichmann, J., Woltering, T. and Mutel, V. Neuropharmacology 39 (2000) 1700.
[23] Malherbe, P., Kratzeisen, C., Lundstrom, K., Richards, J.G., Faull, R.L.M. and Mutel, V. Mol. Brain Res. 67 (1999) 201.
[24] Monastyrskaia, K., Goepfert, F., Hochstrasser, R., Acuna, G., Leighton, J., Pink, J.R. and Lundstrom, K. J. Receptor & Sign. Transd. Res. 19 (1999) 687.

G protein-coupled receptor expression in *Halobacterium salinarum*

Ann M. Winter-Vann, Lynell Martinez, Cynthia Bartus,
Agata Levay, and George J. Turner[*]

Department of Physiology & Biophysics, P.O. Box 016430, University of Miami School of
Medicine, Miami, FL 33101 USA. [*]*Email: gturner@miami.edu.*

Introduction

A fundamental problem in the study of transmembrane proteins is the availability of protein, in the quantities required for detailed biochemical and biophysical characterization [see reviews 1,2]. Atomic resolution structures have solved for a handful of membrane proteins, including photoreaction centers [3-6], prostaglandin H2 synthase [7], porins of various species [8,9], cytochrome *c* oxidase [10], bacteriorhodopsin [11,12] a potassium channel [13], and bovine rhodopsin [14]. The structural characterization of these proteins was achieved, in part, because they were naturally available at very high expression levels.

Members of the G protein-coupled receptor super family (GPCR) are ubiquitous components in the cellular pathways for transmembrane signaling [15]. All GPCRs bind specific ligands and initiate cellular signaling cascades via interaction with heterotrimeric, guanine regulatory "G" proteins [16]. Even though the predicted structural and topographical features of GPCRs are conserved [17], the functions of these seven transmembrane spanning proteins are remarkably diverse.

Except for the visual pigments structural characterization of the GPCR has been severely hampered by their low endogenous expression levels of GPCRs and the inability of protein expression systems to generate large quantities of these proteins [18]. In general, the amounts of receptors obtained are low (0.01 to 100 pmol per milligram of membrane protein) and the goal of obtaining the quantities of GPCR required for biochemical and biophysical studies has not been achieved. Much effort has been expended to express GPCRs in heterologous systems. Low level expression of β_2-adrenergic, serotonin 5-HT$_{1A}$ and rat α_{2B} receptors has been achieved in mammalian

S.R.Kiihne and H.J.M.deGroot (ed.), Perspectives on Solid State NMR in Biology, 141-159.

cell lines, yeast, and *E. coli* [19-22]. A homologous system has been developed in *Rhodopseudomonas viridis* to generate mutants of the photosynthetic reaction center [23]. Homologous overproduction of the *Dictyostelium discoideum* cAMP receptor and the *Saccharomyces cerevisiae* α mating factor mating factor generated 70,000-300,000 receptor copies per cell [24,25]. While there are few successful examples of heterologous expression in yeast, recent reports of Bacteriorhodopsin over-expression [26], Human α2-adrenergic receptor [27] and the plant plasma membrane H^+ ATPase [28] are encouraging. Transient expression of rhodopsin and β2-adrenergic receptor has been achieved in *Xenopus* oocytes [29,30]. However, the yeast and *Xenopus* systems are difficult to scale up to the amounts of protein required for biochemical, structural, and therapeutic evaluation. Even homologous overproduction of *E. coli* membrane proteins has been problematic [31-33] although it has been achieved for ATP synthase and fumarate reductase [34,35]. Eukaryotic viral expression systems have been successfully used to express significant quantities of functional GPCRs. The Semliki forest virus system has fairly high levels of receptors (10-50pmol/mg protein) [36,37,38, and p.131-139 of this volume]. The baculovirus-insect cell system has yielded receptors at levels ranging from 0.02-5pmol/mg protein [39-42]. Khorana and colleagues have achieved stable expression of milligram quantities of synthetic *bovine opsin* gene in HEK293 cells [43] but as yet this strategy has not been successfully applied to other GPCRs.

Some of the most dramatic membrane protein expression successes result from attempts at over-expression of BR both homologously in *H. salinarum* and heterologously in *E. coli*. Under the control of the *bop* gene promoter, numerous Bacteriorhodopsin mutants and the low abundance BR homologues (halorhodopsin, HR, and sensory-rhodopsin, SR) have been over-expressed in yields of 1-30mgs/L protein in *H. salinarum* [44-48, and Turner, unpublished].

H. salinarum is an extremely halophilic member of the Archaebacteria, a phylogenetically distinct group of prokaryotic organisms that are as distantly related to the Eubacteria as they are to the Eukaryotes [49]. *H. salinarum* is cultured in the laboratory utilizing techniques similar to those employed for *E. coli* [50,51]. *H. salinarum* contains a specialized brightly colored membrane (PM) which consists of a complex of one protein (bacterio-opsin, Bop) along with the chromophore retinal in a 1:1 stoichiometric ratio [52]. This complex was named Bacteriorhodopsin since, at the time of its discovery, the only other known retinal binding protein was the visual pigment rhodopsin [53]. The formation of highly ordered 2-dimensional crystalline patches of BR in the purple membrane has facilitated purification and structural analyses.

BR is a member of the seven-transmembrane protein family. This family of proteins possesses a conserved secondary structure characterized by seven alpha-helical segments, which traverse the membrane [17,54]. A recent atomic resolution structure of bovine rhodopsin has been solved and highlights the similarities and differences between BR and some GPCR [14]. Notably rhodopsin exhibited larger and more organized extramembrane regions, consistent with the role of GPCRs in activating G proteins. Spectroscopic analyses demonstrate that conformational changes occur on

GPCR activation, likely driven by changes in the orientation of the transbilayer helices. Even though the orientation of the rhodopsin transmembrane helices, in the ground state, is different from that of BR and the spectra of rhodopsin phototransients resemble those of the BR intermediate states. Over-expression of the GPCR in *H. salinarum* may also form 2-dimensional crystalline patches *in vivo*. Indeed, the low abundance homologue HR, when over-expressed under control of the *bop* promoter formed 2D in vivo lattices [47]. In addition the yeast pheromone receptor (Ste2) when expressed by a similar strategy also formed a high-density membrane fraction [55].

Exploitation of the BR expression potential is feasible since the regulation of bacterio-*op*sin (*bop*) gene expression is one of the best characterized genetic systems in the halophilic Archaea. Knowledge of *bop* gene expression has been advanced by the development of a polyethylene glycol-mediated transformation system [56] along with the development of plasmid shuttle vectors which confer resistance to mevinolin (an HMG-CoA reductase inhibitor [57] or novobiocin (a DNA gyrase inhibitor [58]). The *bop* gene has been cloned [59], and factors that regulate its expression (e.g., oxygen tension, light intensity, and retinal synthesis) have been investigated [50,51]. Betlach and coworkers [50,60] identified the *bop* gene promoter and putative regulatory factor binding sites. DasSarma and coworkers have recently completed mutagenic analysis of the bop gene transcriptional promoter and found it to be in the class of very strong promoters [61]. Translational regulation of bacterio-opsin also occurs, in a Signal Recognition Particle dependent manner [62]. Recently, Krebs and coworkers have experimentally verified co-translational membrane insertion of Bop [63]. The efficiency of coupling between translation and membrane insertion may prevent accumulation of misfolded protein and is anticipated to be a critical advantage in using this system for the high level expression of heterologous proteins.

In this report, we describe our efforts to generate a novel protein expression system that makes use of *bop* gene transcriptional and translational control elements. As representatives of the GPCR superfamily the platelet activating receptor (PAFR) human muscarinic acetylcholine receptor (M1), the bovine rhodopsin (RHO), and the human angiotensin type 1 (AT1) receptor coding regions were selected for attempted expression in *H. salinarum*. Oesterhelt and colleagues recently used a similar strategy to express functional β2 adrenergic receptors in *H. salinarum* [70].

Methods

Bacterial strains. The Escherichia coli K12 strain used was DH5α (F⁻, recA1, endA1, gyrA96, thi-1, hsdR17 (K^+_k, m^+_k), supE44, 1⁻). The H. salinarum strain was L33 (Vac⁻, BR⁻, Rub⁻; [64]).

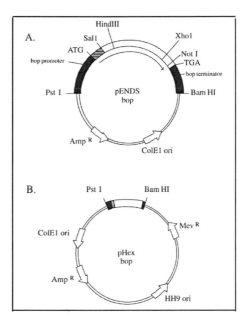

Figure 1. Schematic representation of pENDS and pHex vectors. (A) pENDS bop·contains the 792 bp bop gene plus 370 bp upstream and 87 bp downstream of the bop gene coding region. (B) The pHex bop expression vector contains the 1.25 Kb PstI/BamHI fragment derived from pENDS bop, plus the ColEI origin of replication and ampicillin resistance gene (amp^R) and the H. salinarum pHH9 origin of replication and mevinolin resistance gene (mev^R).

Media and Growth Conditions. All salts and chemicals were reagent grade from standard biochemical supply houses. Bacteriological peptone was from Oxoid (Unipath LTD., Hampshire, England); yeast extract tryptone, Difco peptone, and bacto-agar were from Difco Laboratories (Detroit, MI). Complex E. coli medium was yeast extract tryptone. Preparation of H. salinarum media has been described previously [50,51,65]. H. salinarum transformation was accomplished essentially as described [55,56]. Selective H. salinarum media (broth and plates) were supplemented with 10-25 µM mevinolin (gift of Merck, Sharp and Dohme, Rathaway, NJ).

Individual transformants were isolated from selective solid media and used to establish 1 liter nonselective cultures, in 3 liter Fernbach flasks, as described [55]. Growth was monitored by absorbance at 660 nm, in a Beckman DU50 spectrophotometer.

Vector Constructions. T4 DNA ligase and various restriction endonucleases were purchased from New England Biolabs (Beverly, MA) or from Boehringer Mannheim (Indianapolis, IN). AmpliTaq DNA Polymerase was obtained from Perkin-Elmer (Norwalk, CT). [γ-^{32}P] CTP, and enhanced chemiluminescence (ECL) Western detection reagents were from Amersham, Arlington Heights, IL. Custom oligo- deoxy-

TABLE I: 5' Fusion Constructions

bop or bop fusion construction	nucleotide sequence AMINO ACID SEQUENCE														
nucleotide number:	34														69
[a]bop wt	gta	tcg	cag	gcc	cag	atc	acc	gga	cgt	…… …… ……	ccg	gag	tgg	atc	tgg
bop A,B	gta	tcg	cag	gcc	cag	atc	acc	ggT[b]	cgA	CA'G GCG CTG	ccg	gag	tgg	atc	tgg
	[d]V	S	Q	A	Q	I	T	G	R	Q A L	P	E	W	T	W
pafr A,B	gta	tcg	cag	gcc	cag	atc	acc	ggT	cgA	cag[e] gag cca	cat	gac	tcc	tcc	cac
	V	S	Q	A	Q	I	T	G	R	Q E P	H	D	S	S	H
ml A,B	gta	tcg	cag	gcc	cag	atc	acc	ggT	cgA	CAG GCG CTG	atg	aac	act	tca	gcc
	V	S	Q	A	Q	I	T	G	R	Q A L	M	N	T	S	A
atl A,B	gta	tcg	cag	gcc	cag	atc	acc	ggT	cgA	cac gac gat	tcc	ccc	aag	gct	ggc
	V	S	Q	A	Q	I	T	G	R	Q D D	C	P	K	A	G
rho A	gta	tcg	cag	gcc	cag	atc	acc	ggT	cgA	CAG GCG CTG	aac	ggt	acc	gaa	ggc
	V	S	Q	A	Q	I	T	G	R	Q A L	N	G	T	E	G
nucleotide number:	400												162		
[a]bop wt	gtg	aaa	ggg	atg	ggc	gtc	…	…	tcg	gac	cca	gat	gca	aag	aaa
bop C,D	gtg	aaa	ggg	atg	ggc	gtc	A'AG CTT		tcg	gac	cca	gat	gca	aag	aaa
	V	K	G	M	G	V	K L		S	D	P	D	A	K	K
pafr C,D	gtg	aaa	ggg	atg	ggc	gtc	AAG CTT		gtc	tta	gcc	cgc	ctg	tac	cct
	V	K	G	M	G	V	K L		V	F	A	R	L	Y	P
ml C,D	gtg	aaa	ggg	atg	ggc	gtc	AAG CTT		ttc	aag	gtc	aac	acg	gag	ctc
	V	K	G	M	G	V	K L		F	K	V	N	T	E	L
atl B,D	gtg	aaa	ggg	atg	ggc	gtc	AAG CTT		aag	ctg	aag	act	gtg	gcc	agc
	V	K	G	M	G	V	K L		K	L	K	T	V	A	S

[a] modifications/fusions at the 5' of the bop gene
[b,c,f] capitals denote modifications made to introduce DNA restriction sites SalI, AlwNI and HindIII, respectively
[d] single letter amino acid code
[e] italics denote heterologous gene sequences

nucleotide primers were purchased from Gibco (Life Technologies, Rockville, MD). Oligo-directed mutagenesis was performed with the Transformer Site-Directed Mutagenesis kit, Clontech Laboratories, Inc. (Palo Alto, CA). DNA sequencing was accomplished by Sequenase, United States Biochemical (Cleveland. OH). RNA was extracted by the RNAzol (Cinna/Biotecx Laboratories, Friendswood, TX) or the Qiagen RNeasy Total RNA Kit (Chatsworth, CA). Digoxigenin labeled RNA probes were synthesized in vitro using an SP6 polymerase riboprobe kit (Amersham, MA). The Genius Nonradioactive Northern kit was from Boehringer Mannheim. Electrophoresis grade agarose was from FMC Corporation (Rockland, ME). Nucleotide gel blots utilized Nylon Hybond™ membranes (Amersham, MA).

pENDS Vectors.   A series of cloning vectors (designated pENDS, Fig. 1a) were constructed which fused protein coding sequences to *bop* gene transcription and translation control elements [55]. The amino acids comprising the Bop N-terminus constitute a signal sequence [62], and this coding region was included as the N-terminal coding region of all fusions. *SalI* was chosen as the 5' cloning site. Since it is unknown whether additional sequences within the *bop* gene coding region contribute to expression, fusions were constructed which also included the entire coding region for the BR first transmembrane helix (Fig. 2 and Table I and II).

A unique *Not*I restriction site exists within the C-terminal coding region of the *bop* gene and was used as the 3' cloning site for the heterologous constructions. Inclusion of the downstream *Not*I/*Bam*HI fragment generates a fusion which codes for the last five Bop C-terminal residues and includes the *bop* gene translation stop codon and major transcription terminator [59]. Additional modifications to the downstream fusion involved inclusion of coding sequences that serve to "tag" the heterologous gene products [55]. The tags are short peptide sequences which generate an antigenic epitope (HA tag) and allow immunogenic detection or 6 sequential histidines (6His) useful for affinity purification of the expressed protein. The tag sequences replaced DNA encoding the Bop carboxyl terminal aspartate, while retaining the *bop* gene translational stop and transcriptional termination sequences. The pENDS vectors constructed are described in Tables I, II, and illustrated in Fig. 2. Since it is unknown whether additional sequences within the *bop* gene coding region contribute to expression, gene fusions were designed that also included the entire *bop* gene extramembranous C-terminal coding region (Fig. 2 and Table I and II).

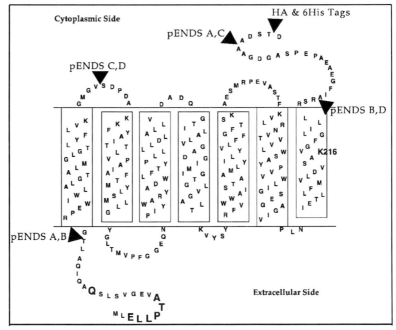

Figure 2. Secondary structure representation of Bop showing the location of the coding region fusion sites for the pENDS:A,B,C,D vectors. The signal sequence and first amino acid (glutamine) of the mature Bop N-terminus are in bold. Also highlighted is lysine 216, the retinal chromophore attachment site in BR. The sites of the Bop-GPCR fusions are indicated by arrowheads and pENDS nomenclature. Inserted or altered DNA sequences and corresponding amino acids are listed in Tables I and II.

TABLE II
3' Fusion Constructions

bop or *bop* fusion construction	nucleotide sequence AMINO ACID SEQUENCE															
nucleotide number:	757															789
	|															|
[a]wt. bop	gcc	gcc	gac	ggc	gcg	gcc	gcg	acc	agc	……… …… …… …… …… …... …… …… ……						gac tga
bop A,C	gcc	ggc	gac	ggc	gcg	gcc	gcg	acc	agc	…… …… …… …… …… …… …… ……						gaT tga
	[b]A	G	D	G	A	A	A	T	S							D tga
bop:HA	gcc	ggc	gac	ggc	gcg	gcc	gcg	acc	agc	TAC CCA TAC GAC GTC CCA GAC TAC GCT tga						
	A	G	D	G	A	A	A	T	S	Y P Y D V P D Y A						
bop:6His	gcc	ggc	gac	ggc	gcg	gcc	gcg	acc	agc	CAT CAC CAT CAC CAT CAC …… …… …… tga						
	A	G	D	G	A	A	A	T	S	H H H H H H						
pafr A,C	[d]tcc	ctc	aaa	aat	[c]GCG	GCC	GCg	acc	agc	TAC CCA TAC GAC GTC CCA GAC TAC GCT tga						
	S	L	K	N	A	A	A	T	S	Y P Y D V P D Y A						
mI A,C	tcc	cgc	caa	tgc	GCG	GCC	GCg	acc	agc	TAC CCA TAC GAC GTC CCA GAC TAC GCT tga						
	S	R	Q	C	A	A	A	T	S	Y P Y D V P D Y A						
atl A,C	ttt	gag	gtg	gag	GCG	GCC	GCg	acc	agc	TAC CCA TAC GAC GTC CCA GAC TAC GCT tga						
	F	E	V	E	A	A	A	T	S	Y P Y D V P D Y A						
rho A	gcg	ctc	gcc	taa	GCG	GCC	GCg	acc	agc	…… …… …… …… …… …… …… ……						gaT tga
	A	P	A	OCH												

nucleotide number:	400															448
	|															|
[f]wt bop	ggc	ttc	ggg	ctc	atc	ctc	ctg	cgc	agt	cgt	gcg	atc	ttc	ggc	gaa	gcc
	G	F	G	L	I	L	L	R	S	R	A	I	F	G	E	A
bop B,D	ggc	ttc	ggg	ctc	atc	ctc	ctg	cgc	TCG	[g]AgG	gcg	atc	ttc	ggc	gaa	gcc
	G	F	G	L	I	L	L	R	S	R	A	I	F	G	E	A
pafr B,D	cct	gtt	atc	tac	tgt	ttc	ctc	cgc	TCG	AgG	gcg	atc	ttc	ggc	gaa	gcc
	P	V	I	Y	C	F	L	R	S	R	A	I	F	G	E	A
mI B,D	tgc	tac	gca	ctc	tgc	aac	aaa	cgc	TCG	AgG	gcg	atc	ttc	ggc	gaa	gcc
	C	Y	A	L	C	N	K	R	S	R	A	I	F	G	E	A
atl B,D	ctg	ttc	tac	ggc	ttt	ctg	ggg	cgc	TCG	AgG	gcg	atc	ttc	ggc	gaa	gcc
	L	F	Y	G	F	L	G	R	S	R	A	I	F	G	E	A

[a] modifications/fusions at the 3' of the bop gene.
[b] single letter amino acid code.
[c] C-terminal tag sequences.
[d] italics denote heterologous gene sequences.
[c,g] capitols denote modifications made to introduce DNA restriction sites *Not*I and *Xho*I respectively.
[f] modifications/fusions in the Bop C-terminal extrahelical region. These *bop-gpcr* fusions also contain the C-terminal HA tag.

Heterologous genes were engineered to contain unique 5' *Sal*I or *Hind*III and 3' *Xho*I or *Not*I DNA restriction sites by oligo-directed mutagenesis or PCR. Restriction mapping and DNA sequencing confirmed all nucleotide changes. The coding regions of the following genes were used to test the expression system: i) the rat angiotensin receptor, type 1 (*atl*), ii) the human acetylcholine muscarinic receptor, type M1 (*mI*) iii) the human platelet activating receptor (*pafr*) and iv) bovine rhodopsin (*rho*). The nucleotide changes required to introduce nucleotide restriction sites in the coding regions of these genes are described in Table I and II.

pHex Vector Constructions. H. salinarum vectors (pHex, Fig. 1b) were constructed for attempted expression [55,65,66]. *Pst*I/*Bam*HI fragments containing the fusion constructs were isolated from pENDS vectors and subcloned into pHex vectors.

Table III.

Relative mRNA and protein levels for pHex bop·A-D constructs

Vector construction	Percent mRNA*	Percent protein‡
bop·A	100	100
bop·B	87	95
bop·C	93	96
bop·D	97	100
bop·A HA	13	10
bop·A 6His	85	80

Total RNA was isolated from 1 milliliter of early stationary H. salinarum cells. 5 μg of each RNA sample was used for Northern analysis and quantified against a serial dilution of bop·A mRNA. Data are indicated as a percentage of bop·A mRNA levels. ‡Protein concentrations were determined spectrophotometrically from the absorbance at 568 nm for whole membranes. Protein levels are indicated as a percentage of the Bop A protein level.

Characterization of Expression. Nucleotide modifications of the *bop* gene coding and noncoding sequences have the potential to disrupt transgenic Bop expression. Therefore, *bop* genes with altered sequences were transformed into *H. salinarum* strain L33 (Bop⁻) and tested for their effect on BR accumulation. The chromophoric properties of BR provide a simple phenotypic screen of BR levels based on color. Colonies or cell pellets with high levels of BR are intensely purple, while colonies with low levels of BR are white. Following transformation into *H. salinarum,* colonies were plated onto solid selective media, inspected for purple color and compared with BR⁺ and Bop⁻ *H. salinarum* strains. BR containing membranes were isolated as described previously [55,67]. BR concentration in the PM was quantified using a Perkin Elmer λ18 spectrophotometer and LabSphere RSA 150 millimeter light scattering attachment (ε^{BR}_{568} = 62,700 M^{-1} cm^{-1}; [68]].

All pHex vectors containing *bop-gpcr* constructions were transformed into *H. salinarum* strain L33 and mev^r colonies were isolated and purified. DNA and mRNA were harvested and characterized by gel blot analyses as described [50,55,65]. The blots were probed with: a) [γ -^{32}P] CTP or UTP random priming of PCR fragments encompassing the *bop* gene or the coding regions of the *gpcr* genes of interest, b) a digoxigenin labeled dUTP *in vitro* transcribed RNA probe generated from a *Kpn*I/*Not*I *bop* gene internal fragment, subcloned into the pSPT19-vector, c) a digoxigenin labeled *Alw*NI/*Not*I DNA fragment containing the coding region of interest, or d) a digoxigenin labeled *Pst*I/*Alw*NI fragment containing the *bop* gene upstream noncoding region and the presequence coding region. Chemiluminescent detection was accomplished following the protocols of Engler-Blum [69].

Total RNA concentrations were determined by absorbance at 260 nm. For *bop-gpcr* gene fusions, 1 to 10 µg of total RNA was used for electrophoretic analysis. Gel loads and RNA quality was checked on duplicate gels by ethidium bromide staining of the 16s and 23s rRNAs. The *bop* gene presequence probe was used to assay the heterologous gene fusion mRNA levels since the presequence coding region was included in all constructions. *bop-gpcr* gene mRNA levels were determined by comparison with the *bop* A mRNA (Fig. 4). Blots were exposed for various times to allow comparison of heterologous gene mRNA with *bop* mRNA concentrations within the saturation limits of the film. Blots were digitized and quantified on an IS2000 Digital Imaging System and software package (Alpha Innotech Corporation, San Leandro, CA).

Bop-GPCR protein expression was characterized by 4-20% Tris-glycine PAGE analysis (NOVEX, San Diego, CA). Western analysis of the AT1, M1, and PAFR receptor proteins relied on the C-terminal HA tagged sequences and the monoclonal antibody 12CA5 (BabCO, Berkeley, CA). Western analysis of Rho relied on a monoclonal antibody derived from peptides representative of N-terminal extrahelical sequences contained in the Rho-A construction (a gift of Prof. Heidi Hamm, Northwestern University, Chicago, Il).

Results and Discussion

We constructed cloning vectors, designated pENDS·A,B,C,D, for linking *bop* gene sequences to coding regions of Eukaryotic *gpcr's* (Fig. 1a, Fig. 2, and Table I and II). We found it necessary to make changes in *bop* gene sequences in order to allow construction of *bop-gpcr* gene fusions. We tested each modification for its effect on *bop* gene expression. The nucleotide modifications required to insert unique DNA restriction sites into the different pENDS vectors had little effect on *bop* expression (Table III). Where possible, the changes required to generate unique restriction sites were introduced as silent mutations; they did not alter the Bop amino acid sequence.

H. salinarum cultures transformed with pHex vectors containing the *bop-gpcr* gene fusions were grown under conditions known to induce high level BR production [65]. DNA gel blot analysis showed the presence of pHex vectors containing the *bop-gpcr* fusion constructions in all cultures analyzed (data not shown). This indicates that there are no significant DNA restriction barriers to overcome upon introducing Eukaryotic genes into *H. salinarum*. RNA gel blot analysis identified full-length transcript for all gene fusions analyzed. Cross reactivity of RNA from the background strain L33 and the heterologous gene probes was not observed (Fig. 3, 4). Only in the case of the *m1* gene RNA probe was cross reactivity with the *bop* gene mRNA observed (Fig. 3). A component of the *m1* mRNA was significantly smaller than the full length, and its origin is unknown at this time. The component appears regardless of denaturation conditions and stringency of blot probing and washing. Either the *m1* mRNA is inherently fragile and some shearing of the full length mRNA occurs during purification, there is a stable folded conformer of the *m1* mRNA, or there

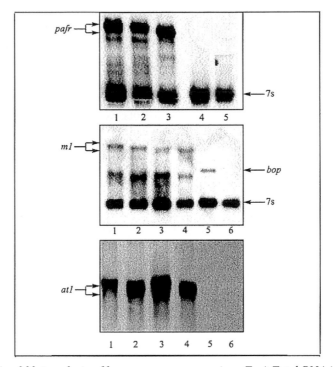

Figure 3. RNA gel blot analysis of bop-gpcr gene expression. Top) Total RNA isolated from H. salinarum strains expressing pafrHA: constructions. bop-pafr mRNA was identified with a random primed SalI-NotI restriction fragment isolated from the bop-pafr cDNA. 5mg of total RNA was loaded for each lane. Lane 1 pafrHA A, Lane 2 pafrHA C, Lane 3 pafrHA D, Lanes 4 and 5 contained RNA isolated from strains L33 and bopA respectively. The bracketed arrow (top left) indicates the full length bop-pafr mRNA's. The gel was simultaneously probed for the H. salinarum 7s RNA (bottom right arrow) to control for gel loading. Middle) Total RNA gel isolated from strains expressing pHex m1HA constructions. bop-m1 mRNA was identified with a random primed AlwNI-NotI restriction fragment isolated from the bop-m1 cDNA. 5mg of total RNA was loaded for each lane. Lane 1 m1HA A, Lane 2 m1HA B, Lane 3 m1HA C, Lane 4 m1HA D, Lanes 5 and 6 contained RNA isolated from strains L33 and bopA respectively. The bracketed arrow (top left) indicates the full length bop-m1 mRNA's. The gel was simultaneously probed for the H. salinarum 7s RNA (bottom band, right) to control for gel loading. The bop mRNA fortuitously cross-reacted with the m1 probe and its size is indicated by the middle arrow (right). Bottom) Total RNA gel isolated from H. salinarum strains expressing pHex at1HA: constructions. bop-at1 mRNA was identified with a random primed SalI-NotI restriction fragment isolated from the bop-at1 cDNA. 5mg of total RNA was loaded for each lane. Lane 1 at1HA A, Lane 2 at1HA B, Lane 3 at1HA C, Lane 4 at1HA D, Lanes 5 and 6 contained RNA isolated from strains L33 and bopA respectively. The bracketed arrow (top left) indicates the full length bop-at1 mRNA's.

exists a cryptic transcription initiation site within the *m1* gene coding region. We have not pursued the origin of this smaller RNA.

In all cases, the levels of accumulation of *bop-gpcr* mRNA was low compared to that for *bop* gene. Fig. 4 illustrates that the levels of *rhoA* mRNA are more than 10 times lower than that observed for the *bop* mRNA. It was possible to detect the *bop-gpcr* and *bop* mRNA on the same gel blot when probed with the *Alw*NI-*Not*I presequence probe as it contained the coding sequence for the Bop signal peptide (contained in constructions). It was necessary to load at least 10 µg of total RNA to see full length mRNA's of *bop-gpcr* fusion constructions whereas 1 µg of total RNA was more than sufficient to identify the *bop* mRNA with this probe.

We initiated the four-fusion strategy to determine if the inclusion of extensive *bop* gene sequences would restore high levels of mRNA accumulation for the *bop-gpcr* mRNA's. We observed that the four-fusion strategy did affect the amounts of mRNA that accumulated but the accumulation was not systematic. As seen in Fig. 3 accumulation of the *bop-at1* mRNAs was highly dependent on the *bop* gene sequences included. Exchanging the entire extramembranous C-terminal coding sequences of the *at1* gene for those of *bop* (construction B) resulted in more *at1* mRNA. Similarly exchanging the entire N-terminal coding sequences through the first helix of the *at1* gene for those of *bop* (construction C) resulted in even more *at1* mRNA. However, the effect of the double swap (construction D) was not additive. It therefor appears that the amounts of full length *bop-gpcr* mRNA that accumulated was combinatorial. mRNA levels dependent on both the *gpcr* gene and the extent of the *bop* gene fusion, but not systematically. Increasing the amount of the *bop* gene contained in the fusion did not necessarily increase *bop-gpcr* mRNA accumulation. Since the levels of mRNA observed for any the *bop-gpcr* gene fusions did not accumulate at high levels there must exist coding region determinants of expression within the *bop* gene itself. Alternatively, the *gpcr* gene coding sequences that have been used may contain inhibitory determinants of high level mRNA accumulation. Since we have now attempted expression of six different *gpcr* coding regions (the current work and ref 54) and low levels of mRNA have been observed for each it is likely that the limiting factor is due to removal of *bop* gene determinants of expression.

To determine the effect of the *bop* gene fusion strategy on protein accumulation whole cell lysates and membrane fractions were examined by SDS-PAGE and Western analyses. The Western analysis of the PAFR, M1 and AT1 fusion proteins relied on inclusion of the C-terminal HA tag (Fig. 2 and Tables II and III). In all cases attempted the Western analysis was negative indicating that the GPCRs were expressed at levels below the detection limits of the 12CA5 antibody, not expressed at all, were unstable in the *H. salinarum* membrane, or that the HA tag itself adversely affected expression. A positive Western signal was observed for membrane fractions isolated from *H. salinarum* expressing the *rhoA* construction (Fig. 4). This construction did not contain the HA tag and detection relied on the monoclonal antibody (gift of Professor Heidi Hamm, Northwestern University). Comparison of the detection intensities allows an estimate of Rho expression levels of 1-5 pmol/mg membrane protein. Visible

A.M.WINTER-VANN, ET AL.

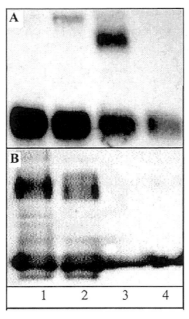

Figure 4. pHex rhoA expression in H. salinarum.   Top panel, RNA gel blot showing relative amounts of full length rhoA and bopA mRNA.   Lane 1, 10 μg of total RNA isolated from strain L33; Lane 2, 10 μg of total RNA isolated from strain expressing pHex rhoA; Lane 3, 1 μg of total RNA isolated from bopA; Lane 4, 1 μg of total RNA isolated from strain L33.   The blot was probed with a random primed PCR probe that represented the coding sequence of the 13 amino acid bop presequence (contained in both the rhoA and bopA constructions).   The blot was simultaneously probed for the H. salinarum 7s RNA (lower band) to control for gel loading. Bottom panel, Western analysis of RhoA expression. Rod Outer Segments (Lanes 1 and 2, 30.0 pmol) and whole membrane fractions (~5mg membrane protein loaded) from two H. salinarum isolates expressing RhoA (Lanes 3 and 4) were electrophoresed on 12% polyacrylamide gels and detected with a rhodopsin monoclonal antibody.   At least half of the Rho control ran as dimers (upper bands in lane 1 and 2).

spectroscopy of the membrane fraction did not indicate that the expressed Rho coupled with the retinal isomers synthesized by *H. salinarum* or that the expression levels were below our spectral sensitivity. However, upon incubation of 10 mgs of total membranes with 40mM 11-cis retinal a small Rho absorbance was observed (personal communication, Dr. Drake Mitchell, National Institutes of Health, Bethesda, Md). The Rho absorbance was detected from a difference spectrum of membrane before and after incubation with retinal and was too small to allow quantification. The detection limit for this spectrophotometer and the 40μl cuvette used is ~5pmol Rhodopsin. The spectra indicate that Rho was expressed in a form competent to incorporate its ligand. Oesterhelt and colleagues demonstrated that *H. salinarum* cells expressing the β2 adrenergic receptor had a rank order of potency for numerous receptor ligands that mirrored that of receptors expressed in mammalian cell lines [70]. It appears then that functional GPCR can be expressed in *H. salinarum*.

We had previously observed that C-terminal HA tagged GPCRs could be identified in a Western analysis [55]. Since there are in general no antibodies available for use in the Western analysis of GPCR expression we have to rely on an "epitope tagging" strategy. We have also observed that inclusion of tag sequences in the C-terminal coding region of the *bop* gene affected the levels of *bop* mRNA and BR [54]. Inclusion of short peptide coding sequences (HA and 6His tags) in the pENDS A vector decreased BR accumulation by as much as ten-fold (Figs. 5). Northern analysis revealed full length mRNA for *bop*·HA and *bop*·6His but levels were reduced similar to that observed for BR levels (Fig. 5 and Table III). Inclusion of the HA tag in the *bop-gpcr* fusion strategy may similarly have a dominant and negative effect on the potential for mRNA and protein accumulation. We are currently assessing this potential by a systematic study of C-terminal coding region introduced in a number of the *gpcr* genes used in the current study and by generation of polyclonal antibodies to peptides representative of the Bop N-terminal amino acids contained in all fusion constructions. Our preliminary data indicate the N-terminal polyclonals are useful in the Western analysis of numerous BR controls (G. Turner, unpublished).

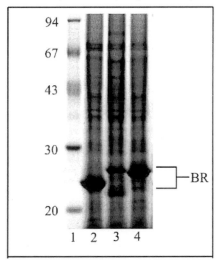

Figure 5. SDS-PAGE analysis showing the effect of bop gene C-terminal coding region modifications on BR accumulation. Lanes 1 are molecular size markers; sizes are shown to the right in kDa. Lanes 2-4 contain equivalent total membrane proteins loads for BR·A, BR·HA, and BR·6His, respectively. The brackets to the lower right indicate the size and relative abundance of the BR constructions.

Conclusions

Over-expression of transmembrane proteins is critical in order to study their roles in the transmission of external signals and solutes across membrane barriers. In general, the transgenic expression of membrane proteins, in amounts sufficient for purification and structural analysis, has met with limited success [1,2]. Currently the highest levels of expression appear to have been achieved with the Semliki forest virus and stably transfected CHO cell systems [36-38,43]. In this report we describe our continuing efforts to develop an expression system with the potential to yield multiple milligram quantities of membrane proteins. The system we are pursuing uses the promoter and regulatory factors responsible for the naturally high level expression of BR in the halophilic archaeon *H. salinarum*. We constructed a series of expression vectors that fuse coding regions of heterologous proteins with DNA sequences known to be required for the expression of the *bop* gene [55]. We have also made extensive *bop-gpcr* gene fusions to determine their effect on expression levels.

To simplify the use of this system we altered *bop* gene sequences to create convenient cloning sites and to stabilize *bop-gpcr* gene transcription and/or translation products. We also introduced specific sequences to tag gene products for the purposes of identification or purification. The N-terminal coding region of the *bop* gene is known to contain numerous determinants of expression [59,61,63]. Therefore, upstream sequence modifications and fusions were designed to minimize changes in regions involved in polymerase binding, initiation of transcription, the ribosome binding site, or the N-terminal signal sequence [55, and this work].

In contrast, alterations in the C-terminal extrahelical coding region variably affected *bop* gene expression (Table III). Replacing the BR C-terminal aspartate with small peptide coding sequences had a negative effect on *bop* gene expression. The HA coding sequence had a more severe effect than the 6HIS coding sequence (Figs. 5,6). The correspondence between mRNA and protein accumulation for BopA, *bop*·HA, and *bop*·6His indicates that the effect is at the level of transcription (Fig 6 and Table III). The perturbations observed were not due to alteration of the *bop* gene transcriptional terminator, which has been mapped to untranslated sequence nearly 50 bp downstream of the location of the C-terminal tags [59] and is intact in our constructions. Other investigators observed that truncation of the C-terminus of sensory rhodopsin I (SRI) resulted in a dramatic increase in SRI expression from the *bop* gene promoter [47]. In this case the truncation did not effect *sensory-opsin I* mRNA levels, and a role for the deleted C-terminal amino acids in translational regulation was hypothesized.

Our downstream cloning strategy utilized an endogenous *Not*I restriction site, located 12 bp upstream of the *bop* gene translational stop signal. Cloned heterologous gene sequences may adversely effect the potential for expression due to close proximity with the sequence sensitive downstream coding region of the *bop* gene, as evidenced by the HA and 6His tags. The C-terminally tagged Bop constructions indicate that levels of BR accumulation are tightly coupled to the levels of mRNA accumulation. For the

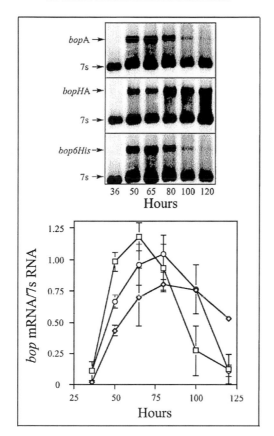

Figure 6. Top panel: RNA gel blot analysis showing the time course of accumulation of bacterio-opsin gene mRNA. 2µg of total RNA was loaded for each lane. Each gel blot was probed with an identical volume of a probe mixture containing random primed SalI-NotI restriction fragment isolated from the bop cDNA and the 7s RNA. The 7s RNA allowed normalization to total RNA loaded and allowed quantitative comparison between blots. Top, Time course of induction of accumulation for bopA mRNA. Middle, Time course of induction for bop6His mRNA. Bottom, time course of induction for bopHA mRNA. Bottom panel: Quantitation of the bop gene mRNA gel blots. The intensities of the bop mRNA and 7s RNA were determined by densitometry. Ratios of the are plotted against the time of cell harvest and RNA isolation. The open squares are bopA/7S ratios, the open circles are bop6His/7s ratios, and the open diamonds are bopHA/7s ratios. This analysis quantifies the differences in timing and amount of bop gene mRNA accumulation resulting from the C-terminal coding region tag sequences.

BR tagged examples the rank order of both protein and mRNA accumulation were BR>BR6His>>BRHA. The tight linkage between mRNA and protein levels observed suggests that optimization of *bop-gpcr* fusion mRNA levels is a critical component for the success of this expression system. One of the critical steps in such optimization involves be the removal of the HA tag sequences, work currently underway. The data presented indicate that there may be previously unappreciated determinant of high level *bop* gene expression in the C-terminal coding region. A logical extension is that other determinants of high level expression may reside elsewhere in the *bop* gene coding region.

We tested the effect of extensive coding region swaps on *gpcr* expression. No single *bop-gpcr* fusion was sufficient to achieve wild-type *bop* gene levels of mRNA accumulation. Additionally, the effects of the different fusions on individual *gpcr* coding region expression are not systematic and their effects are different on the various *gpcr* coding regions. In all cases we were able to detect full length *bop-gpcr* mRNA indicating that transcription of the fusion genes is feasible. The HA tagged yeast Ste2 receptor and Rho have been successfully expressed, at low levels [55, and current work]. The success in expressing two (of six) GPCR constructions, and the success of expressing functional β2 adrenergic receptors by Oesterhelt and colleagues [70] verifies our expression strategy.

Except where necessary to expedite cloning, the codon usage for these non-Archaeal genes was not modified. It is unclear whether the low level of expression was due to codon usage bias (*H. salinarum* coding regions are ~70% GC), C-terminal transcriptional effects, mRNA stability, or other unknown factors. We have previously shown that high level expression of heterologous gene is feasible with this system by expressing an *E.coli* soluble protein fused to the BR C-terminus (BR·AT) [55,66]. The high level expression of BR·AT established that the AT-rich *E. coli* codon usage bias was not a critically limiting factor [55,66]. In this case the entire *bop* gene coding region was present which supports our emerging hypothesis that critical determinants of expression may reside in the *bop* gene coding region. Identifying these determinants and their transfer into heterologous gene coding regions may be a critical requirement to achieve our expression goals and is a focus of our continuing efforts.

Acknowledgements

Support is acknowledged in the form of an American Heart Association Grant-in-Aid (AHA664871), the National Science Foundation (MCB-9817140) and a Johnson & Johnson Focused Giving Program Grant.

References

[1]   Grisshammer, R., and Tate, C. G. Quarterly Rev. Biophys. 28 (1995) 318.
[2]   Tate, C. G., and Grisshammer, R. Trends in Biotechnology, 14 (1996) 426.
[3]   Allen, J. P., Feher, G., Yeates, T. O., Komiya, H., and Rees, D. C. Proc. Natl. Acad. Sci. USA 84 (1987) 5730.
[4]   Deisenhoffer, J., Epp, O., Miki, K., Huber, R., and Michel, H. Nature 318 (1985) 618.
[5]   Chang, C-H., El-Kabbani, O., Tiede, D., Norris, J., and Schiffer, M. Biochemistry 30 (1991) 5352.

[6]   McDermott, G., Prince, S. M., Freer, A. A., Hawthornthwaite, A. M., Papiz, M. Z., Cogdell, R. J., and Isaacs, N. W. Nature 374 (1995) 517.

[7]   Picot, D., Loll, P. J., and Garavito, R. M. Nature 367 (1994) 243.

[8]   Weiss, M. S., Dreusch, A., Schiltz, U., Nestel, U., Welte, W., Weckesser, J., and Schulz, G. E. FEBS Lett. 280 (1991) 379.

[9]   Cowan, S. W., Schirmer, T., Rummel, G., Steiert, X., Ghosh, R., Pauptit, R. A., Jansonius, J. N., and Rosenbusch, J. P. Nature 358 (1992) 727.

[10]   Ostermeier, C., Iwata, S., Ludwig, B., and Michel, H. Nature Structural Biology 2 (1995) 842.

[11]   Grigorieff, N., Ceska, T. A., Downing, K. H., Baldwin, J. M., and Henderson, R. J. Mol. Biol. 259 (1996) 393.

[12]   Pebay-Peyroula, E., Rummel, G., Rosenbusch, J. .,P and Landau, E. M. Science 277 (1997) 1676.

[13]   Doyle, D.A., Cabral, J.M., Pfuetzner, R.A., Kuo, A., Gulbis, J.M., Cohen, S.L., Chait, B.T., and MacKinnon, R. Science 280 (1998) 69.

[14]   Palczweski, K., Kumasaka, T., Hori, T. Behnke, C.A., Motoshima, H., Fox, B.A., Le Trong, I., Teller, D.C., Okada, T., Stenkamp, R.E., Yamamoto, M., Miyano, M. Science 289 (2000) 239.

[15]   Hargrave, P. Current Opinion in Structural Biology 1 (1991) 575.

[16]   Gilman, A.G.A. Ann. Rev. Biochem. 56 (1987) 615-649.

[17]   Baldwin, J.M., Schertler, G.F.X. and Unger, V.M. J. Mol. Biol. 272 (1997) 144.

[18]   Schertler, G.F.X. Current Opinion in Structural Biology 2 (1992) 534.

[19]   Bertin, B., Freissmuth, M., Breyer, R.M., Schutz, W., Strosberg, A.D. and Marullo, S. J Biol. Chem. 267 (1992) 8200.

[20]   Xia, Y., Chhajlani, V. and Wikberg, J.E.S. Eur. J. Pharm. 246 (1993) 129.

[21]   Breyer, R.M., Strosberg, A.D., and Guillet, J.G. EMBO J. 9 (1990) 2679.

[22]   King, K., Dohlman, H.G., Thorner, J., Caron, M.G. and Lefkowitz, R.J. Science 250 (1990) 121.

[23]   Laussermair, E. and Oesterhelt, D. (1992) EMBO J. 11 (1992) 777.

[24]   Johnson R.L., Vaughan, R.A., Caternia, M.J. VanHaastert, P.J.M., and Devreotes, P.N. Biochemistry. 30 (1991) 6982.

[25]   Blumer, K.J., Reneke, J.E., Thorner J. J. Biol. Chem. 263 (1988) 10836.

[26]   Hildebrandt, V., Ramezani-Rad, M., Swida, U., Wrede, P., Grzesiek, S., Primke, M., and Buldt, G. FEBS Lett. 243 (1989) 137.

[27]   King K., Dohlman, H.G., Thorner, J., Caron, M.G., and Lefkowitz, R.J. Science. 250 (1990) 121.

[28]   Villalba, J.M., Palmgren, M.G., Berberian, G.E., Ferguson, C. and Serrano, R. J. Bio. Chem. 267 (1992) 12341.

[29]   Khorana, H.G. J. Biol Chem. 267 (1992) 1.

[30]   Savarese, T.M. and Fraser, C.M. Biochemistry. 283 (1992) 1.

[31]   Saier, M.H. Jr., Werner, P.K., and Muller, M. Microbiol Rev. 53 (1989) 333.

[32]   van Weeghel, R.P., Keck, W., and Robillard, G.T. Proc. Natl. Acad. Sci. U.S.A. 158 (1990) 590.

[33]   Eckert, B., and Beck, C.F. J. Bacteriol. 171 (1989) 3557.

[34]   von Meyenburg, K., Jorgensen, B.B., and van Deurs, B. EMBO J. 3 (1984) 1791.

[35]   Weiner, J.H., Lemire, B.D., Elmes, M.L., Bradley, R.D., and Scraba, D.G. J. Bacteriol. 158 (1984) 590.

[36] Malherbe, P., Kratzeisen, C., Lundstrom, K., Richards, J.G., Faull, R.L., and Mutal, V. Brain Res. Mol. Brain Res. 67 (1999) 201.
[37] Scheer, A., Bjorklof, K., Cotecchia, S., and Lundstrom, K. J. Recept. Signal Transduct. Res. 19 (1999) 369.
[38] Lundstrom, K., Michel, A., Blasey, H., Bernard, A.R., Hovius, R., Vogel, H., and Surprenant, A. J. Recept. Signal Transduct. Res. 17 (1997) 1115.
[39] Marheineke, K., Lenhard, T., Haase, W., Beckers, T., Michel, H., and Reilander, H. Cell. Mol. Neurobiol. 18 (1998) 509.
[40] Nowell, K.W., Pettit, D.A., Cabral, W.A., Zimmerman, H.W., Abood, M.E., and Cabral, G.A. Biochem. Pharmacol. 55 (1998) 1893
[41] Blaukat, A. Herzer, K., Schroeder, C., Bachmann, M., Nash, N., Muller-Esterl, W. Biochemistry 38 (1999) 1300.
[42] Brys, R., Josson, K., Csatelli, M.P., Jurzak, M., Lijnen, P., Gommeren, W., and Leysen, J.E. Mol. Pharmacol. 57 (2000) 1132.
[43] Reeves, P.J., Thurmond, R.L., and Khorana, H.G. . Proc. Natl. Acad. Sci. USA. 93 (1996) 11487.
[44] Turner, G.J., Miercke, L., Thorgeirsson, T., Kliger, D., Betlach, M., and Stroud, R. Biochemistry 32: (1993) 1332.
[45] Ni, B.F., Chang, M., Duschl, A., Lanyi, J. and Needleman, R. Gene 90 (1990) 169.
[46] Krebs, M.P., Hauss, T., Heyn, M.P., RajBhandary, U.L., and Khorana, H.G. Proc. Natl. Acad. Sci. USA. 88 (1991) 859.
[47] Heymann, J.A.W., Haveika, W.A., and Oesterhlet, D. Molecular Microbiology 7 (1993) 623.
[48] Ferrando-May, E., Brustmann, B., and Oesterhelt, D. Molecular Microbiology 9 (1993) 943.
[49] Woese, C., and Fox, G. Proc. Natl. Acad. Sci. USA. 74 (1977) 5088.
[50] Betlach, M. C. and Shand, R. F. General and Applied Aspects of Halophilic Microorganisms Plenum Press, New York and London, (1991) 259.
[51] Robb F.T.(ed. in chief), DasSarma, S. and Fleischmann, E.M.(ed.), Cold Spring Harbor Laboratory Press, Plainview, New York, (1995).
[52] Oesterhelt, D., and Stoeckenius, W. Proc. Natl. Acad. Sci. USA. 70 (1973) 2853.
[53] Oesterhelt, D., and Stoeckenius, W. Nature 233 (1971) 149.
[54] Henderson, R., Baldwin, J.M., Ceska, T.A., Zemlin, F., Beckman, E., and K.H. Downing J. Mol. Biol. 213 (1990) 899.
[55] Turner, G.J., Reusch, R., Winter-Vann, A., and Betlach, M.B. Protein Expression and Purification 17 (1999) 325.
[56] Cline, S.W., Lam, W.L., Charlebois, R.L., Schalkwyk, L.C., and Doolittle, W. Canadian Journal of Microbiology 35 (1989) 148.
[57] Blaseio, U. and Pfeifer, F. Proc. Natl. Acad. Sci. U.S.A. 87 (1990) 6772.
[58] Holmes, M.L. and Dyall-Smith, M.L. J. Bacteriol. 173 (1991) 642.
[59] Dunn, R., McCoy, J., Simsek, M., Majumdar, A., Chang, S. H., RajBhandary, U. L., and Khorana, H.G. Proc. Natl. Acad. Sci. USA 78 (1981) 6744.
[60] Gropp, F., and Betlach, M. C. Molecular Microbiology 16: (1995) 357.
[61] Baliga, N.S. and DasSarma, S. J. Bacteriol. 181 (1999) 2513.
[62] Gropp, R., Gropp, F., and Betlach, M. (1992) Proc. Natl. Acad. Sci. USA 89 (1992) 1204.
[63] Dale, H., and Krebs, M.P. J. Biol. Chem. 271 (1999) 22693.

[64]   Wagner, G., Oesterhelt, D., Krippahl, G., and Lanyi, J. FEBS Letters 131 (1983) 341.
[65]   Winter-Vann, A.M., Martinez, L. C., Parker, L., Talbot, J. D., and Turner, G. J. Cancer Research, Therapy and Prevention 8 (1999) 275.
[66]   Turner, G.J., Miercke, L.J.W., Winter-Vann, A., Schafmeister, C., Mitra, A., Betlach, M.B. and Stroud, R.M. Protein Expression and Purification 17 (1999) 312.
[67]   Oesterhelt, D., and Stoeckenius, W. Methods in Enzymology 31 (1974) 667.
[68]   Rehorek, M., and Heyn, M. P. Biochemistry 18 (1979) 4977.
[69]   Engler-Blum, G., Meie, M., Frank, J., and Muller, G. A. Anal. Biochem. 210 (1993) 235.
[70]   Sohlemann, P., Soppa, J., Oesterhelt, D., and Loshe, M. J. Naunyn-Schmiedeberg's Arch Pharmacol. 355 (1997) 150.

Magnetic resonance microscopy for studying the development of chicken and mouse embryos

**Robert E. Poelmann[a], Bianca Hogers[a], Huub J. M. de Groot[b],
Cees Erkelens[b], Dieter Gross[c], Adriana C. Gittenberger-de Groot[a]**

*[a]Department of Anatomy and Embryology, PO Box 9602, 2300 RC, Leiden, The Netherlands,
[b]Leiden Institute of Chemistry, Leiden, The Netherlands,
[c]Bruker Analytic GmbH, Rheinstetten, Germany.*

Introduction

In this new era of transgenic mouse models, the need to evaluate the effects of gene manipulation during embryonic development is increasing. Traditional embryological studies are invasive, sacrificing embryos before processing in order to reveal specific information like morphology, histology, antigen distribution, gene expression patterns, or physiological parameters. Each objective requires a specific method, excluding the use of one specimen for multiple questions, while for temporal information each method has to be repeated on several specimens during development. The non-invasive character of magnetic resonance microscopy (MRM) allows the study of subsequent stages of normal development in a single embryo *in utero* over extended periods. In addition, MRM offers the possibility of studying the onset and course of a malformation during development. MRM has proven to be a powerful tool for fixed embryos [1] and living chicken embryos *in ovo* [2]. Recently, Smith and coworkers [3] managed to visualize and follow living rat embryos *in utero* in a 2.0T MR microscope by a 3D projection encoding technique with a total scanning time of 27 min. The objectives of the present study are to visualize fixed chicken embryos and living mouse embryos *in utero*. Imaging of the embryo requires very high resolution in combination with excellent contrast. To achieve this, we used moderate to ultra high magnetic fields of 7.0 and 17.6T for the chicken material and 7.0T for the mouse embryos *in utero* and explored various fast imaging sequences. Fast imaging is necessary to avoid artifacts from embryonic movements that cannot be controlled by the researcher.

S.R.Kiihne and H.J.M.deGroot (eds.), Perspectives on Solid State NMR in Biology, 161-167.

Methods

Chicken embryos. Fertilized White Leghorn eggs (*Gallus domesticus*) were incubated at 37°C and 60-70% relative humidity for 3-5 days, until stages 15, 24 or 29, according to the age determination criteria of Hamburger and Hamilton [4].

For fixation and mounting of the embryos, the technique of Smith and coworkers[5] was employed. In short: warm (37°C) phosphate-buffered saline was perfused into the omphalomesenteric vein with a glass needle, followed by a fixative perfusion (2% glutaraldehyde/1% formaldehyde in phosphate buffer). Subsequently, a gadolinium contrast agent (Bovine Serum Albumine - Diethylene Triamine Pentaacetic Acetate - Gadolinium chloride (1mM) mixed with 5% gelatin) was injected. Embryos were immersion-fixed at 4°C to solidify the gelatin and transferred to plastic containers filled with proton-free perfluoro-polyether Fomblin (Ausimont Inc., Bollate, Italy). The size of the containers was chosen to be as small as possible, immobilizing the embryo without disturbing its 3D configuration.

Mouse embryos. CPB-S mice were mated overnight and checked for vaginal plugs the next morning. Mice carrying embryos of 13, 16, or 17 days of development were used. The mother mouse was connected to an inhalation anesthesia unit (N_2O/O_2/isoflurane). The anesthetized mouse was mounted in a probe with an internal diameter of 38 mm. Cardiac and respiratory motion were checked by an ECG and a fiber-optic method [6], while gating of these signals was achieved by a Physiogard SM 785 NMR trigger unit (Bruker) equipped with a light-diode transmitter/detector system. The body temperature was secured by tubes filled with circulating warm water. The mouse was inserted head-up into the resonator. The experiments were performed in compliance with the *Guide for the Care and use of Laboratory Animals* (NIH 85-23).

Magnetic Resonance Microscopy. Chicken embryos were imaged using either a vertical 300 MHz (7.0T), 150 mm bore magnet (AVANCE console) or a vertical 750 MHz (17.6T), 89 mm bore magnet (DSX-750 console, both from Bruker Analytic, Rheinstetten, Germany). The 750 MHz spectrometer was equipped with a 89 mm bore shim unit (inner diameter 72 mm). Both systems were connected to microimaging accessories, gradient systems of 100 G/cm and Micro2.5 probes with exchangeable rf-coils. We used a 4 mm solenoid coil for the stage 15 embryo, a 10 mm birdcage coil for the stage 24 embryo and a 15 mm birdcage coil for the stage 29 embryo.

Each session started with a multislice orthogonal gradient echo sequence for positioning and selection of the desired region for subsequent experiments. Data sets were obtained by a 3D spin echo imaging method. For the stage 15 embryo, the matrix size (MTX) was 256.128.128, the echotime (TE) 4 ms, and the repetition time (TR) 100 ms. A signal average of 2 was used for each phase encoding step. For the other embryos (stage 24 and 29) MTX was 256.256.256, TE 6 ms, TR 200 ms, while we used 4 averages. The field of view (FOV) was adjusted to match the size of the embryo and the total experiment time (T) was adapted accordingly. For stage 15, the FOV was 8x4x4 mm and T was 53 min, for stage 24 the FOV was 10x10x10 mm and T was 14h33min, and for stage 29 the FOV was 14x14x14 mm and T was 14h33min (300 MHz) and 14h46min (750 MHz), respectively.

In vivo imaging of mouse embryos was performed as follows. The magnetic field homogeneity was optimized by shimming. Each session began with a multislice orthogonal gradient echo sequence for position determination and selection of the desired region for subsequent analyses.

Diffusion spin echo imaging was used to obtain anatomical images of living mouse embryos at 14 days of pregnancy. With a FOV of 40 mm and MTX of 256.256.256, we obtained a resolution (R) of 156 μm. TR for pulse sequences was 2000 ms, and TE was 30.3 ms, with two averages for each phase encoding step, resulting in T = 17min4sec.

To reduce T, we explored an imaging method known as the multi-slice RARE method [7]. This method employs a single excitation step followed by the collection of multiple phase encoded echoes. For this purpose, mice of 13 and 17 days of pregnancy were imaged. With a FOV of 50 mm (16 slices) and an MTX of 256. 256. 256, we obtained an R of 195 μm. TR was 3910 ms for the 13-day-old embryo, and 5000 ms for the 17-day-old embryo, while TE was 11 ms. We used a RARE factor of 16, with four averages for each phase encoding for the youngest embryos and two averages for the older embryos, resulting in T = 2min10sec for the 13-day-old embryo and T = 5min31sec for the 17-day-old embryo.

To simultaneously obtain functional information, two imaging methods were combined, RARE and very fast gradient echo (VFGE). As reference proton image, we selected one slice from the RARE data set. Subsequently, gradient echo imaging with short TR (30 ms) and TE (4.3 ms) and a 90° pulse angle was performed on the same slice, with T = 30 sec.

Data acquisition and processing were performed with ParaVision (Bruker, Rheinstetten, Germany) running on a Silicon Graphics O2 workstation with the Irix 6.5.3 operating system.

Results and Discussion

Figure 1a shows a fixed chicken embryo stage 24, imaged at 300 MHz, revealing the complexity of the vascular system. An intravascular contrast agent was used to enhance images of the vascular tree. Maximum intensity projections were obtained from the datasets, resulting in full 3D images. The oldest embryo (stage 29) has also been imaged in the 750 MHz magnet (Fig. 1b). We used the same parameter settings for TR (200 ms) and TE (6 ms) to demonstrate the improvement of image quality rather than shortening the total scanning time. SNR, defined as (mean signal intensity)/(standard deviation of the noise), was determined at various positions in the 3D images taken at 300 MHz and at 750 MHz. The higher magnetic field improved SNR by a factor of 3 in volumes without contrast agent and by a factor of 3.4 in blood vessels with contrast agent. The difference in these factors may be caused by differential saturation of the magnetization

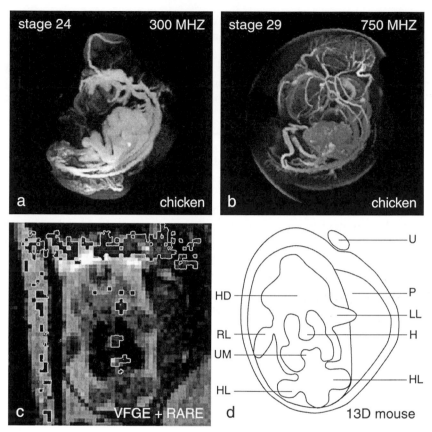

Fig 1a represents a fixed, stage 24 chicken embryo after injection of Gadolinium into the vascular bed. A detailed image of all the major vessels has been obtained in a 300 MHz NMR.. Fig 1b. Image of a stage 29 embryo at 750 MHz. This image shows considerably more detail compared with the image of the same embryo in a 300 MHz NMR (not shown). Fig 1c. A doubly imaged embryo using consecutively VFGE and RARE compressed into one figure presents rich tissue contrast semiquantitatively combined with flowing blood (white contours). Fig 1d. represente a simplified sceme of fig 1c. H heart, HD head, HL hindlimbs, LL left forelimb, P placenta, RL right forelimb, U uterine artery, UM umbilicus.

and different T2 relaxation times in the tissues with and without contrast agent. Further experiments must be performed to determine the precise T1 and T2 relaxation times for the tissues, blood vessels with contrast agent and field strength. The contrast between the blood vessels and the other tissues is determined as C = (Difference of the mean intensities of tissue with and without contrast agent) /(mean intensity of tissue without contrast agent) [8]. The gain in contrast ($C_{750\ MHz}/C_{300\ MHz}$) is approximately 20 percent. Faster repetition times might increase the contrast but with the additional risk of reducing the overall SNR.

The 750 MHz image (Fig.1b) shows the same pattern as the 300 MHz magnet but with a better CNR. CNR is determined as the difference of the SNR values from tissue with and without contrast agent [8] at various positions in the 3D images at 300 MHz and 750 MHz respectively. A gain of a factor of 3.5 is realized. Small vessels that are beyond detection in the 300 MHz image, for instance in the brain and eyes, can be observed with the 750 MHz MRM. There is also more detail within the heart. In particular, details of the atria, ventricles and outflow tract that are hardly visible in the 300 MHz image form clearly separate structures in the ultra high field image. This indicates that susceptibility effects do not cause a loss of structural information in the images at the very high field of 17.6 T from fixed chicken embryo samples. The influence of the better SNR and CNR resulted in more structural information [9].

To our knowledge, this report concerns the first imaging of embryonic stages using ultra high magnetic field equipment, showing considerable improvement at high field relative to current practice [5]. The better performance at 750 MHz is mainly attributed to the improved sensitivity in the high field. Since we kept the experimental parameters constant, the SNR improves with the applied field strength. In addition, the resolving power in high field is not limited by susceptibility artifacts with the parameter settings used. Imaging with the 750 MHz magnet is thus promising for the future as younger embryos that have smaller blood vessel diameters can be studied in detail.

Although there is an improvement of vessel size recognition in the 750 MHZ instrumentation, we were not able to visualize in any stage the very small diameter vessels. This is primarily attributed to the high viscosity of the contrast agent. It appears necessary to optimize the contrast agent properties to find a balance between maximal perfusion of the smallest embryonic vessels, requiring low viscosity, and maximum gadolinium fixation inside the vessels, where a more viscous agent is prefered. For the experiments described here a 5% gelatin concentration was used, which is already half of the gelatin concentration used elsewhere [5].

Diffusion spin echo is an adequate method for visualizing contrast between organs of the mother mouse, such as uterine tissues, and of embryos depicting brain vesicles, eyes, thorax, and extremities. Solid tissues, like the brain, were depicted bright, while fluid-filled structures, like the eye and gut were darker grey. Regions that contain moving fluid, like blood in the placenta, liver and heart, and even amniotic fluid in the amniotic cavity, were black. Although diffusion spin echo acquisitions resulted in clear anatomical images, the total scanning time of 17 min was undesirably long.

The RARE sequence was the ultimate method for embryonic contrast (Fig 1c). This method resulted in high-contrast T2 weighted proton images of the embryos and the placenta. As this is a very fast method, it is possible to increase the resolution at the expense of the registration time by increasing the number of averages per slice. This improves SNR within a reasonable amount of time. From the 17-day-old mouse, a detailed image of the placenta, the umbilical vessels within the umbilical cord, the head of the embryo including brain vesicles, heart, liver, extremities, spine, ribs, and even the dorsal aorta and the carotid artery was obtained. In contrast to the SE diffusion method, the amniotic fluid presents white with RARE. We obtained comparable results in a 13-day pregnancy.

Finally, we succeeded in demonstrating intraembryonic blood flow in a 13-day pregnant mouse without the use of contrast agent. We used Very Fast Gradient Echo (VFGE) on one slice selected from the RARE dataset (Fig 1c, d). The resulting image was representative of fluid flow perpendicular to the image plane. With postprocessing this image was superimposed on the RARE image, allowing us to visualize blood flow through the heart and dorsal aorta of the embryo, and also through the placenta and uterine arteries in a semi-quantitative way.

The MRM sequences presented here allow for contrast-rich images of living mouse embryos *in utero*. To visualize heart development in very small objects, i.e. embryos from approximately 3 mm to 2 cm, the highest possible resolution is necessary. Imaging at 7.0 Tesla was adequate to discern the heart as a separate structure from 13 days of development onwards, but we expect that higher magnetic fields are necessary to obtain more cardiac anatomical details.

By combining the RARE method for obtaining a contrast-rich image of an embryo with VFGE, we were able to obtain functional information in addition to the morphological image. The fast repetition time combined with a 90° excitation pulse caused a saturation of the magnetization from the static spins in the selected slice. As a consequence, the image intensity of the static tissue disappeared completely, but blood flowing into the selected imaging plane replaces the saturated magnetization. This results in a net enhancement of the signal of the flowing blood above background levels. By choosing different planes at the level of the embryonic heart, the direction and velocity of blood flow can be determined. In case of abnormal heart development, as is pertinent in many transgenic mouse models, this method might reveal functional cardiac abnormalities well before anatomic defects can be detected.

Another advantage of these fast imaging procedures is that all embryos within the womb can be examined individually in a reasonable time. After scanning the complete abdominal region of the mother, very small FOV's can be selected for the individual embryos resulting in higher spatial resolution. Imaging 10 embryos, each with a resolution of 40-50 μm, may take approximately 15-20 minutes.

Conclusion

It is demonstrated that, with appropriate acquisition times in ultra high magnetic field, imaging of embryos is feasible with both good contrast and resolution . Further improvement of the resolution can be expected in the near future by technological developments. High field imaging of a living embryo within reasonable experiment times will make it possible to study longitudinally the development of the cardiovascular system and the ontogeny of cardiovascular anomalies in one single affected embryo. In conclusion, fast imaging sequences currently result in respectable morphological images of living mouse embryos *in utero*. Development of MRM technology in the near future will further improve the quality and time of acquisition. It must be emphasized that the value of using MRM is the fact that it can complement other analytical tools, since MRM is non-invasive and non-deleterious, whereas subsequently the sample can be processed for other studies. This is an important advantage when the object is a very rare specimen, like a transgenic animal model.

Acknowledgements

This research was supported by a Grant-in-aid from the Leiden University. The ultra high field spectrometer is financed in part by the Commission of the European Communities through demonstration project BIO4-CT97-2101. We thank Jan Lens for photographical assistance.

References

[1]   Huang, G.Y., Wessels, A., Smith, B.R., Linask, K.K., Ewart, J.L., Lo, C.W., Dev. Biol. 198 (1998) 32.
[2]   Effman, E.L., Johnson, G.A., Smith, B.R., Talbott, G.A., Cofer, G., Teratology 38 (1988) 59.
[3]   Smith BR, Shattuck MD, Hedlund LW, Johnson GA. Magnet Resonance Med 39 (1998) 673.
[4]   Hamburger, V., Hamilton, H.L., J Morphol 88 (1951) 49.
[5]   Smith BR, Johnson GA, Groman EV, Linney E., Proc Natl Acad Sci U S A 91(1994) 3530.
[6]   Wilson, S.J., Brereton, I.M., Hockings, P., Roffmann, W., Doddrell, D.M., Magn Reson Imaging 11(1993) 1027.
[7]   Henning, J., Nauerth, A., Friedburg, H., Magnet Reson Med 3 (1986) 823.
[8]   Wehrli KU. In Wehrli, F,W., Shaw, D., Kneeland, J.B. (eds) Biomedical Magnetic Resonance Imaging. Principles, Methodology, and Applications. 1988.
[9]   Smith, B.R., Linney, E., Huff, D.S., Johnson, G.A., Comput Med Imaging Graph 20 (1996) 483.

SECTION IV:

Applications in Membrane Proteins and Peptides

MAS NMR on a uniformly [^{13}C, ^{15}N] labeled LH2 light-harvesting complex from *Rhodopseudomonas acidophila* 10050 at ultra-high magnetic fields

T. A. Egorova-Zachernyuk[a], J. Hollander[a], N. Fraser[b], P. Gast[c], A.J. Hoff[c], R. Cogdell[b], H.J.M. de Groot[a] & M. Baldus[a§]

[a]*Gorlaeus Laboratories and* [c]*Huygens Laboratories, Leiden University, 2333 CC Leiden, The Netherlands.* [b]*Division of Biochemistry and Molecular Biology, University of Glasgow G12 8QQ, United Kingdom.*
§) *Present address: Max-Planck-Institute for Biophysical Chemistry, Am Fassberg 11, 37077 Göttingen, Germany*

Introduction

Solid State Nuclear Magnetic resonance (SSNMR) spectroscopy is considered to be one of the tools for structure determinations of membrane proteins, and this technique, along with X-ray crystallography, will play an important role in structural genomics projects. The goal of structural genomics is the determination of the 3-D structure of all human proteins or of the complete sets of proteins in particular functional classes, such as enzymes or cell-surface receptors. As of today, 19 different structures of polytopic membrane proteins from inner membranes of bacteria and mitochondria and from eukaryotic membranes, 16 structures of membrane proteins from the outer membrane of gram negative bacteria and related membrane proteins, and 4 structures of the monotopic membrane proteins that are only inserted into the membrane have been determined by crystallographic methods [1]. At the same time, about 2000 structures of water soluble proteins have been determined. SS NMR is a tool for structure determination of the membrane proteins that can not be crystallized and for the structure-functional studies of the membrane proteins of known structure. Within this context it is important to get assignment data for solid state NMR studies.

Structure determination of immobilized globular and membrane proteins using SS NMR techniques [2,3] has made considerable progress [4-6]. So far, stable isotope

S.R.Kiihne and H.J.M.deGroot (eds.), Perspectives on Solid State NMR in Biology, 171-183.

labeling in conjunction with Magic Angle Spinning [7] or preferential sample orientation techniques [6] have been most successful in achieving the sensitivity and resolution necessary for biological SSNMR studies. MAS-based methods have been successfully applied to determine internuclear distances [8] and torsion angles [9] well beyond the resolution obtainable by X-ray crystallography. Most of these studies involved singly- or doubly labeled compounds customized for the biophysical problem under study. As recently demonstrated in uniformly labeled ^{13}C networks, [10, 11] MAS studies involving multiply labeled biomolecules may permit structural investigations with greater flexibility and may profit from residue specific chemical shift information obtained in the liquid-state [12]. Prerequisites for structure determination of partially [13] or uniformly labeled peptides and proteins [14-16] include hetero- and homonuclear assignment techniques. For both aspects, a variety of pulse schemes have been proposed and recently employed [17,18] for nearly complete assignment of a uniformly labeled SH3 domain containing 62 residues and many of the resonances of the membrane spanning part of light-harvesting complex LH2 from the purple non-sulphur photosynthetic bacterium *Rhodopseudomonas acidophila* 10050 [15,19]. Partial assignments were also reported for Ubiquitin [13, 20] and BPTI [14]. These systems are suitable candidates for establishing MAS based protocols that determine conformations of peptide ligands bound to their membrane-protein receptor target [20] or structural constraints in globular proteins [21]. The number of residues in these peptides also compares favorably to membrane-spanning or surface-bound peptides studied recently in oriented lipid bilayers [6, 22]. It can thus be inferred that MAS-based correlation techniques could also be used to study entire membrane-protein topologies or subsections thereof.

In this contribution we are making a next step towards membrane protein structure elucidation under MAS and show that heteronuclear (^{13}C,^{15}N) correlations of significant resolution can be obtained in a uniformly labeled membrane-protein environment. A system has been chosen for examining and improving NMR techniques for applications in multi-labeled membrane-proteins. Uniformly (^{13}C,^{15}N) labeled LH2 light-harvesting complex from the purple non-sulphur photosynthetic bacterium *Rhodopseudomonas acidophila* 10050 strain [23] plays an important role in photosynthesis. In this complex, the initial event in bacterial photosythesis starts: a photon is absorbed by the light-harvesting antenna system and energy is rapidly and efficiently transfered to the reaction center where charge separation takes place. To our knowledge, the present system (about 150 kD) represents the largest integral membrane-protein complex investigated by NMR so far. We here report results obtained using a novel wide-bore 750 MHz NMR instrument offering increased resolution and sensitivity compared to standard-size systems. With this work an attempt to achieve a sequential assignment and type-specific amino acid assignment in a membrane protein complex is made.

The crystal asymmetric unit contains three protomer complexes. These complexes consist of an α - (53 amino acid residues) and a β (41 amino acid residues) apoprotein, three bacteriochlorophyll *a* molecules, one rhodopin-glucoside carotenoid,

α: M[1]NQGKIWTVV[10]N**PAIGIPALL**[20]**GSVTVIAIL**V[30]**HLAIL**SHTTW[40]F**P**AYWQGGVK[50]KAA
    N-terminal   Membrane-spanning   C-terminal

β: ATLTAEQSEE[10]LH**KYVIDGTR**[20]**VFLGLALVAH**[30]**FLAFSA**TPWL[40]H
    N-terminal   Membrane-spanning   C-terminal portion

Fig. 1. Amino acid sequence for the α and β protein subunits of Rhodopseudomonas acidophila 10050 B800-B850 complex. Membrane spanning residues are indicated in bold. Tentatively assigned or identified residues in this work are underlined.

and a β - octylglucoside detergent molecule. Both proteins have been sequenced [24] (Fig. 1). The transmembrane helices of nine α -apoproteins are packed side by side to form a hollow cylinder of 18 Å radius. The nine helical β -apoproteins are arranged radially with the α -apoprotein to form an outer cylinder of 34 Å. [23]

Material and Methods

[U-^{13}C,^{15}N] enriched LH2 light-harvesting complex was obtained by growing purple photosynthetic bacteria *Rhodopseudomonas acidopila* 10050 anaerobically in light at 30 °C on a well defined medium containing [U-^{13}C, ^{15}N] algae hydrolysate and [U-^{13}C, ^{15}N] labeled (NH$_4$)$_2$- Succinate as the sole nitrogen and carbon sources [19]. LH2 complex was subsequently prepared as described in [25] and was characterized by optical absorption spectroscopy.

NMR experiments were carried out on a DSX 750 wide-bore spectrometer (Bruker, Germany). Triple resonance experiments at ultra-high fields were conducted using a 4 mm (^{1}H,^{13}C,^{15}N) triple-channel MAS probe. Solutions of uniformly labeled LH2 in Tris-HCl/ 2% LDAO buffer, pH 8 were concentrated to a volume of 30 μl amounting to 10 mg of protein in a 4 mm CRAMPS rotor. MAS spinning rates between 8 and 12 kHz were used in the temperature range of –10 °C to –50 °C. Stable sample cooling was achieved using a pressurized N$_2$ heat exchanger / dewar system permitting N$_2$ refill intervals of longer than 20 h.

Results and Discussion

1D CPMAS Experiments. One-dimensional (^{13}C)-CPMAS [26] spectra of [U-^{13}C,^{15}N] labeled LH2 at 750 MHz at different temperatures are shown in Fig. 2. R.f. fields during CP and acquisition were optimized for each temperature. In all cases, signal intensities were optimized using amplitude modulated ^{13}C r.f. fields [27] during the (^{1}H,^{13}C) CP transfer step. The influence of sample temperature upon resolution has been investigated. For most of the resonances, we observe a slight improvement in resolution upon raising the temperature. However a further analysis requires the application of multidimensional correlation techniques.

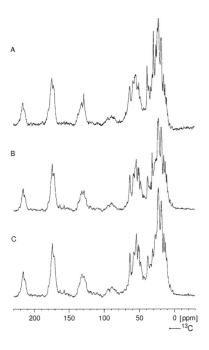

Fig. 2. ^{13}C-CPMAS spectra of [U-^{13}C,^{15}N] LH2 from Rps. acidophila 10050 Recorded at 750 MHz with a MAS frequency of 8 kHz at different temperatures: −10 °C (A), −30 °C (B), −50 °C (C). During acquisition, TPPM [28] proton decoupling at 70 kHz rf field strength was applied.

<u>*2D (NC) Correlation Experiments*</u>. Methods involving only proton and/or carbon resonances during evolution or detection periods are not likely to provide sufficient resolution with MAS at low temperature to resolve entire backbone or sidechain networks. To test the degree of resolution and sensitivity obtainable in the system described above, we focus in the following on 2D-(^{13}C, ^{15}N) correlation experiments. Comparable to the liquid state [29], spectral assignment generally requires hetero- and homonuclear transfer steps that direct polarization from one spin to the next in the polypeptide chain with high efficiency. For applications in uniformly labeled globular or membrane-proteins under MAS, the polarization technique employed should fulfill criteria discussed in [19].

For heteronuclear transfer in polypeptides is discussed in detail how the conventional CP [26] approach can be modified to direct polarization transfer in a chemical shift selective manner [30]. The usefulness of this approach in the context of uniformly labeled globular and partially labeled membrane proteins has already been demonstrated and is discussed in [19]. Band-selective SPECIFIC CP techniques that allow for the observation of NC correlations can be realized by the introduction of a

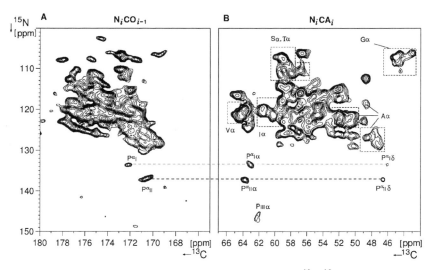

Fig. 3. Heteronuclear (N,C) correlation spectra obtained on [U-^{13}C,^{15}N] LH2. In A and B, the ^{13}C carrier frequency was placed slightly outside CO and C$_\alpha$ region of the ^{13}C spectrum, respectively. Band-selective SPECIFIC CP transfer around the n = 1 was optimized using the experimental setup discussed in [19].. For backbone resonances, correlations of type (N$_i$CO$_{i-1}$, Fig. 3A) and N$_i$Cα$_i$ (Fig. 3B) are expected.

slow amplitude modulation, for instance, on the ^{15}N r.f. channel during a (^{15}N,^{13}C) 2D experiment [19]. Note that the discussed concept achieves polarization transfer and band-selectivity without phase or frequency modulation. It is thus amenable to methodological extensions such as homonuclear decoupling schemes often essential in macroscopically oriented systems [22].

Following the experimental protocol outlined in [19], correlations on [U-^{13}C,^{15}N] LH2 are shown in Fig. 3. After a broadband HN transfer step, band-selective NC transfer (τ(^{15}N,^{13}C) = 2 ms) was optimized to observe prominent NCO (Fig. 3A) and NCA (Fig. 3B) correlations. For Fig. 3, a MAS frequency of 8 kHz and a temperature of –10° C were chosen. TPPM decoupling at 85 kHz was applied in t$_1$ and t$_2$. SPECIFIC CP [30] transfer was established using r.f. fields between 10-25 kHz and optimized for a r.f. carrier frequency positioned slightly outside the CO (Fig. 3A) and the CA (Fig. 3B) resonance region. 128 t$_1$ experiments were performed in a total experiment time of about 4 hours.

In the NCO correlation experiment (Fig. 3A), 10-15 resolved peaks around a relatively broad but structured correlation area at 120 ppm (^{15}N chemical shift) can be identified. As expected, the relatively small chemical shift range of carbonyl resonances leads to significant overlap - even in the 2D experiment. In Fig. 3B, results of an NCA correlation experiment under identical conditions are shown. In line with NMR studies of proteins in solution [12] and recent SSNMR results on globular proteins [14], the increased dispersion in C_α resonances leads to a significant number of resolved resonance lines detectable in the NCA correlation area. Individual peaks with ^{15}N and ^{13}C linewidths of 1.2 ppm and 1 ppm, respectively, are observed. While resolution down to the baseline is only possible for a limited number of these correlations, a total number of about 30-40 peak maxima can be identified. This number is consistent with observation of the membrane-spanning part of the α and β protein subunits (Fig. 1) arranged in high symmetry in the LH2 complex.

Of particular interest are distinct collections of peaks in the NCA spectrum (N_i - $C_{\alpha,i}$) at 42 - 46 ppm and 47 - 49 ppm ^{13}C chemical shift. Based on their characteristic ^{13}C chemical shifts, these correlations can be type-assigned to Gly and Ala residues, respectively. In addition, C_α resonances of Pro residues can be tentatively assigned at 63.2 and 63.6 ppm in Fig. 3B. As expected for a band-selective experiment, we observe for both Pro residues C_δ resonances (Fig. 3B) and correlations of type N_iCO_{i-1} in Fig. 3A. Note that an additional weaker NC_α-type correlation is observed (147 ppm, 62.5 ppm) with a significantly larger line width in the ^{15}N dimension. As discussed in more detailed below, both observations would be consistent with the detection of a third Pro residue in C-terminal portions of the LH2 membrane protein (Fig. 1). Based on

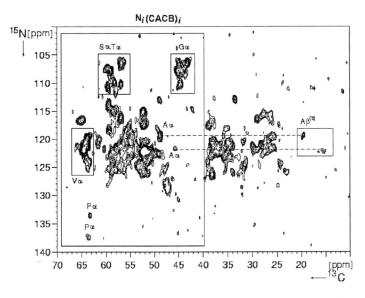

Fig. 4. Heteronuclear (N,C) band-selective double-quantum spectrum obtained on a [U-$^{13}C,^{15}N$] LH2 at 750 MHz with a MAS frequency of 8 kHz at –10 °C. During the evolution and acquisition, TPPM proton decoupling at 71.4 kHz r.f. was used, during the other stages a cw decoupling at 71.4 kHz r.f. was used. For CP ^1H-^{15}N a mixing time of 2 ms, for CP ^{15}N-^{13}C a mixing time of 3 ms, and a spin lock for ^{13}C-^{13}C of 3 ms was applied. 128 t_1 points were accumulated.

chemical shift statistics obtained in the liquid-state [14], possible assignments of Val, Ile, Ser and Thr C_α correlations peaks are also included in Fig. 3B.

Results from liquid-state NMR data [12] are used to identify the C_α resonances areas of the amino acids S,T,G,V,I, and P. In total, three (I-III) correlations involving P residues are detected. For the two stronger sets of signals, additional C_δ and CO resonances of the P residue and its preceding amino acids, respectively, can be identified and are attributed to the membrane-spanning segments in apoprotein α.

2D (NCC) Correlation Experiments. Additional information can be obtained if the NC correlations presented so far are combined with a homonuclear polarization transfer element. In the current context, we have relied on the mechanism of homonuclear spectral spin diffusion ('SD', [31]), where polarization transfer among rare spin carbons is mediated by a strongly coupled proton bath. During the NMR experiment, SD mediated transfer is achieved by insertion of a longitudinal mixing time $\tau(^{13}C,^{13}C)$ after the SPECIFIC CP transfer step and prior to detection in t_2.

Band-selective double-quantum experimental data are shown in Figure 4. Based on chemical shift statistics obtained in the liquid state, assignment of Gly, Ser, Thr , Ala, Val and Pro residues are shown on this figure. The two correlations with ^{15}N chemical shifts of 90 ppm with $R_{20\delta}$ (56.2 ppm) and $R_{20\gamma}$ (31.6) were assigned from this experiment. These correlations are observed as well in Fig.5.

In Fig. 5, we present results of an 2D NCC correlation experiment in which band-selective NCA transfer is followed by a homonuclear SD mixing unit. In the initial rate regime of a homonuclear spin pair, proton-driven spin diffusion rates under MAS are dependent on the size of the MAS spinning frequency ω_R and the isotropic chemical shift difference Δ. In particular, if the rotational resonance condition $\Delta = n\omega_R$ (where n = 1, 2,..) is fulfilled, an enhanced transfer rate is expected [33]. For 8 kHz MAS at 750 MHz, a SD mixing time $\tau(^{13}C,^{13}C)$ of 8 ms is sufficient to observe one-bond correlations. Polarization transfer across two bonds is influenced by the resonance condition discussed above. In general, the combination of broadband HN and band-selective NC transfer also permits correlations within nitrogen containing side chains. This is exemplified in Fig. 6A, where correlations at 90 ppm ^{15}N chemical shift are observed. Inspection of the primary sequence (Fig. 1) predicts only one Arg sidechain correlation at residue 20 of the β apoprotein. The observed ^{15}N chemical shift of 90 ppm agrees well with average chemical shifts observed in soluble proteins and is consistent with model studies involving Arg containing peptides (Baldus et al., unpublished results). We thus assign the two correlations to $R_{20\delta}$ (56.2 ppm) and $R_{20\gamma}$ (31.6 ppm). Moreover, we could also identify the correlations to $R_{20\epsilon}$ and $R_{20\zeta}$ (at 157.3 ppm and 32 ppm respectively, data not shown), and a weaker correlation at 28.5 ppm, indicative for the observation of $R_{20\beta}$.

In principle, the identification of this residue permits a sequential assignment of surrounding residues by comparison either to the $C_\alpha C_\beta C_\gamma$ region of the experiment (Fig. 5B) or to additional homonuclear CC correlation experiments. In a first step, this may

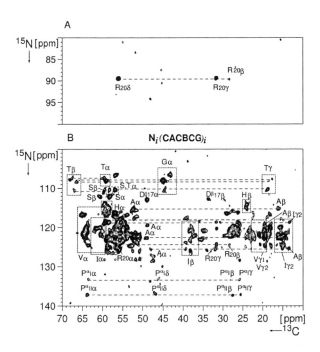

Fig. 5. Combined NC-CC 2D transfer experiment on [U-^{13}C,^{15}N] LH2 that reveals sidechain (Fig. 5A) and backbone (Fig. 5B) N-C-C correlation patterns. As a preparatory step, SPECIFIC CP transfer was used to isolate carbon resonances around 50 ppm. A subsequent spin diffusion time τ(CC) of 8 ms allows for transfer along two additional peptide bonds of type C$_α$C$_β$C$_γ$. In Fig. 5A, sidechain resonances of residue R are identified. In B, various correlation sets in the backbone region of the experiment are indicated by horizontal lines and are discussed in the text.

be accomplished in Fig. 5 by drawing vertical lines from the C$_β$and C$_γ$ correlations observed in Fig. 5A to the C$_α$C$_β$C$_γ$ region of the experiment (Fig 5B). For both ^{13}C shifts, we expect a horizontal correlation representing the C$_α$, C$_β$, and C$_γ$ resonances of R20, tentatively assigned in Fig. 5B. Moreover, the observed correlation provides an internal calibration of the signal intensity of an individual NC or CC correlation in these spectra.

Using the intensity information for the Arg residue, at least 4 Gly correlations are observable. Thr residues are often characterized by C$_β$ chemical shifts well above 60 ppm and C$_γ$ correlations around 20 ppm [12]. As indicated in Fig. 5B, we observe at least three sets of Thr C$_{α-γ}$ correlations for ^{15}N chemical shift values of 107-111 ppm, albeit with significant variations in signal intensities. This number matches the number of membrane-spanning Thr residues. Discrimination between Ser and Thr peaks can be attempted using their different sidechain topologies. As a result, 2-3 Ser correlations can

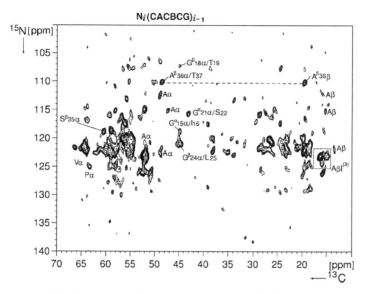

Fig.6. Combined NC-CC 2D transfer experiment on [U-^{13}C,^{15}N] LH2. Experimental setup differs from Fig. 5 by SPECIFIC CP transfer tailored to CO resonances around 178 ppm (under identical parameters as in Fig. 3A). A subsequent spin diffusion time $\tau(CC)$ of 8 ms allows transfer along two additional peptide bonds of the type CO-C$_\alpha$C$_\beta$. The C$_\alpha$C$_\beta$ segment of the experiment containing the largest dispersion is shown. Four correlations involving G residues are assigned. Likewise, A or T containing residue pairs (i,i+1) are indicated in the form S(C$_{\alpha i}$)/A $_{i+1}$, T(C$_{\beta i}$)/V$_{i+1}$ and A(C$_{\alpha,\beta i}$)/T$_{i+1}$. Finally a correlation area for three pairs of the form A$_i$I $_{i+1}$ is indicated.

be identified and are indicated in Fig. 6B. In line with recent observations in immobilized globular proteins [14,15], the observed ^{13}C chemical shifts compare favorably with observations in the liquid-state. On the other hand, the observed ^{15}N chemical shift values deviate by about 5 ppm from the average shifts observed in globular proteins [12].

Additional correlations around and below 20 ppm may result from Ala and Ile residues. In total we expect 14 correlations for both types of amino acids in the membrane-spanning segment. Using information on the characteristic sidechain chemical shift, at least one set of A$_{\alpha,\beta}$ and I$_{\alpha-\gamma2}$ correlations at 115 and 119 (^{15}N) ppm, respectively, can be identified. Possible additional correlation areas as indicated in Fig. 5B can be defined. Comparison between Figs. 5A and 5B indicates that correlations around 15-20 ppm are in general stronger than for the R$_{20\beta}$ peak discussed earlier. This might partially be attributed to the rotational resonance phenomenon mentioned earlier,which favors C$_\alpha$ - C$_\gamma$ polarization transfer across 40 ppm. Moreover, characteristic ^{15}N chemical shifts of Pro residues help to identify two sets of Pro resonances, as expected from the membrane-spanning portion of the protein. Two-bond

C_β and C_γ correlations are clearly visible and agree with the NCA results of Fig. 3B. Finally, areas consistent with average chemical shift values for Val_α, Ile_α, Ile_β and $Val_{\gamma 1}$ correlations are given in Fig. 5.

Interresidue connectivities can be established by combining the information obtained so far with results of an N_i -$(COC_\alpha C_\beta)_{i-1}$ experiment (Fig. 6). In the present context, we will attempt to establish preliminary interresidue contacts for residues containing characteristic chemical shifts to illustrate the principle of sequence-specific assignment under MAS for a membrane protein complex. The limited chemical shift range in the CO part of the spectrum does not, in most cases, permit resolution of individual correlations. However, the NCO transfer serves as a relay step to observe characteristic sidechain topologies of the preceding residue in the polypeptide chain. Since analogous experimental conditions were employed as in Fig. 6, correlation patterns of the form $CO-C_\alpha-C_\beta$ are to be expected. The combination of broadband HN and band-selective NC transfers here also permits correlations of the form $C_\delta-C_\gamma-C_\beta$ and $C_\gamma-C_\beta-C_\alpha$ for Q (Gln) and N (Asn) residues, respectively. However, these amino acids do not occur within the membrane-spanning section of the protein.

In Fig. 6 where the aliphatic region of the $N-CO-C_\alpha-C_\beta$ – type experiment is shown, one of the four interresidue correlations involving G residues can be immediately identified as $(G^{18}T^{19})$ of the β protein subunit. Likewise, tentative assignments for the pairs $G^{15}I^{16}$, $G^{21}S^{22}$ of the α- apoprotein and $G^{24}L^{25}$ of the β-apoprotein in the membrane spanning part are included. Following the discussion presented earlier, correlations observed around 20ppm should predominately arise from Ala residues. Distinct sets of chemical shifts should thus be detectable for the two AT pairs of the protein, namely $A^{36}T^{37}$ at the end of the membrane-spanning part and A^1T^2 at the N-terminus of the β protein segment. One correlation is readily identified in Fig. 6. In line with the results presented so far, it seems probable that stronger signals result from polarization transfer among membrane-spanning residues, and we consequently assign this correlation to the pair $A^{36}T^{37}$. Fig. 6 also contains a tentative assignment of the $A_\alpha^{36}T^{37}$ peak that is consistent with the results of Fig. 5. Characteristic chemical shifts also allow for the assignment of the pair $S^{35}A^{36}$ and for the identification of a T_β correlation at 66 ppm that could be consistently extended to the $V^{23}T^{24}V^{25}$ segment of the α apoprotein. Fig. 6 indicates the occurrence of at least four more AX pairs in close agreement with 3 AI, 2 AL and one AF pair in the membrane. Using the information in Fig. 5, three AI pairs are tentatively identified. Obviously, additional information is needed to completely assign these segments.

As expected, we also observe two proline resonances in the CO portion of Fig.6 (data not shown). For both residues, proton driven polarization transfer to the CACB region of the spectrum, however, is weak. Similar observations were recently made (Pauli *et al.*, unpublished results) in an immobilized globular protein indicative of an additional influence of the proton density at a given residue during SD polarization transfer. Additional experiments e.g. involving coherent homonuclear transfer elements [30,32] might clarify this issue and provide complementary intra –and inter residue information. The protein residues discussed above are summarized in Fig.1. Due to the particular sequence details, the majority of correlations are found in the α segment of the protein.

Conclusions

Two-dimensional backbone and sidechain correlations for a [U-^{13}C,^{15}N] labeled version of the LH2 light-harvesting complex from *Rps. acidophila* 10050 indicate significant resolution at low temperatures and under Magic Angle Spinning. Tentative assignments of some type-specific and sequence-specific amino acid residues of the observed correlations are presented and attributed to the helical segments of the protein, mostly found in the membrane interior.

The data presented so far show that even at low temperatures with MAS, significant resolution in 2D correlation spectra of a uniformly labeled membrane-protein can be obtained. A preliminary analysis of distinct amino residues indicates that significant portions of the membrane-spanning section can be detected. On the other hand, backbone and sidechain correlations involving N and C terminal loop segments of the protein appear to be significantly attenuated.

The observation of an additional Pro residue outside the membrane spanning part of the α apoprotein indicates that residues of defined secondary structure outside the membrane can also be detected. The observation of the alpha-helical membrane-spanning parts of the protein would also be consistent with the limited chemical shift ranges in the recorded NC correlations.

Moreover, experiments e.g. involving coherent (^{13}C,^{13}C) transfer steps, three-dimensional correlation spectroscopy or the incorporation of heteronuclear NC (Spin-Spin) decoupling schemes should enable a more detailed analysis. Although the observed line width is mostly likely dominated by heterogeneous broadening effects, additional line narrowing might result from higher spinning speeds or stronger proton decoupling fields. Additional information could be obtained from multiple-quantum experiments or spectral editing and filtering experiments that simplify the spectral analysis significantly. A subsequent structure elucidation might also profit from long-range proton-proton transfer or torsional constraints. Finally, varying experimental conditions such as temperature, the degree of sample labeling or the choice lipid/detergent environment can be explored to improve the resolution.

Experiments along these lines are currently ongoing in our laboratory and will be reported elsewhere. The data obtained so far hold promise that SSNMR related studies of ligand-binding interactions in membrane-proteins involving small to medium size polypeptides should be feasible. The results obtained so far also indicate that MAS-based correlation experiments might complement structural studies in multiply labeled membrane proteins where macroscopic orientation techniques suffer from decreased spectral resolution. [34] Both approaches contribute to the converging evidence that solid-state NMR can reveal structural information in biophysical systems that are not accessible by other spectroscopic methods at present. It is shown that the sequence-specific assignment of most of the membrane-spanning part of α apoprotein and partially of β apoprotein is feasible by 2D ^{15}N-^{13}C correlations.

Acknowledgments

This project was financed in part by demonstration project (BI04-CT97-2101) of the commission of the European communities. HJMdG is a recipient of a PIONIER award of the Chemical Sciences section of the Nederlandse Organisatie voor Wetenschappelijk Onderzoek (CW-NWO). Support from C. Erkelens during various stages of the project is gratefully acknowledged. We thank D. de Wit for help in microbiological work.

References

[1]   Michel, H. "Crystallization of Membrane Proteins", International Tables for Crystallography, Vol. F., chapter 4.2, in press, (2000). See also http://www.mpibp-frankfurt.mpg.de/michel/public/memprotstruct.html.

[2]   Ernst, R.R., Bodenhausen, G. and Wokaun, A. Principles of Nuclear Magnetic Resonance in One and Two Dimension, Oxford: Claredon Press, 1987.

[3]   Mehring, M. *Principles of High Resolution NMR in solids*, Springer Verlag Berlin.. 1983.

[4]   De Groot, H.J.M. Curr. Opin. Struc. Biol., 10 (2000) 593.

[5]   Smith, S.O., Aschheim, K. and Groesbeek, M. Quart. Rev. Biophys., 29 (1996) 395.

[6]   Griffin, R.G. Nat. Struct. Biol., 5 (1998) 508.

[7]   Andrew, E.R., Bradbury, A. and Eades, R.G., Nature. 182 (1958) 1659.

[8]   Verdegem, P.J.E., Bovee-Geurts, P.H.M, de Grip, W.J., Lugtenburg, J. and de Groot, H.J.M. Biochemistry., 38 (1999) 11316.

[9]   Feng, X., Verdegem, P.J.E., Lee, Y.K., Sandstrom, D., Eden, M., BoveeGeurts, P., deGrip, W.J., Lugtenburg, J., de Groot, H.J.M. and Levitt, M.H. J. Amer. Chem. Soc., 119 (1997) 6853.

[10]   Boender, G.J., Raap, J., Prytulla, S., Oschkinat, H. and de Groot, H.J.M. Chem. Phys. Lett., 237 (1995) 502.

[11]   Egorova - Zachernyuk, T.A., van Rossum, B., Boender, G.J., Franken, E., Ashurst, J., Raap, J., Gast, P., Hoff, A.J., Oschkinat, H. and de Groot, H.J.M. Biochemistry, 36 (1997) 7513.

[12]   Wishart, D.S., Sykes, B.D. and Richards, F.M. J. Mol. Biol., 222 (1991) 311. For a recent statistical compilation of chemical shifts in proteins, see also: Biomolecular NMR Data Bank (BioMagResBank): A Repository for Data from NMR Spectroscopy on Proteins, Peptides, and Nucleic Acids, http://www.bmrb.wisc.edu/

[13]   Hong, M. J. Biomol. NMR, 15 (1999) 1.

[14]   McDermott, A., Polenova, T., Bockmann, A., Zilm, K.W., Paulsen, E.W., Martin, R.W. and Montelione, G.T. J. Biomol. NMR, 16 (2000) 209.

[15]   Pauli J., Baldus M., van Rossum B., de Groot H., Oschkinat H J. Magn Res. 143 (2000) 411.

[16]   Straus, S.K., Bremi, T., and Ernst, R.R. *J. Biomol. NMR,* 12 (1998) 39-50.

[17]   Baldus, M., Geurts, D.G. and Meier, B.H. Solid State NMR 11 (1998) 157.

[18]   Bennett, R.G. and Vega, S. NMR Basic Principles and Progress, (1994) 1.

[19]   Egorova-Zachernyuk T.A., Hollander J., Fraser N., Gast P., Hoff A.J., Cogdell R., de

Groot H.J.M., Baldus M. J. Biomol NMR (2000) Submitted

[20]   Pellegrini, M., and Mierke, D.F. Biopolymers (Peptide Science), 51 (1999) 208.

[21]   Siegal, G., van Duynhoven, J., and Baldus, M. Curr. Opin. Chem. Bio., 3 (1999) 530.

[22]   Opella, S.J., Marassi, F.M., Gesell, J.J., Valente A.P., Kim Y., Oblatt-Montal, M. and Montal, M. Nat. Struct. Biol., 6 (1999) 374.

[23]   McDermott, G., Prince, S.M., Freer, A.A., Hawthorn-Lawless, A.M., Paiz, M.Z., Cogdell, R.J. and Isaacs, N.W. Nature, 374 (1995) 517.

[24]   Zuber, H., and Brunisholz, R.A. *The Chlorophylls* (ed. Scheer, H.) 1993, p. 627.

[25]   Hawthorn, A.M. and Cogdell, R.J. *Bacteriochlorophyll binding proteins in Chlorophylls* (Scheer, H., Ed), CRC Press, Boca Raton, FL, 1991. p. 493.

[26]   Pines, A., Gibby, M.G. and Waugh, J.S. J. Chem. Phys., 59 (1973) 569.

[27]   Metz, G., Wu, X.L. and Smith, S.O. J. Mag. Reson., 110 (1994) 209.

[28]   Bennett, A.E., Rienstra, C.M., Auger, M., Lakshmi, K.V., and Griffin, R.G. J.Chem.Phys., 103 (1995) 6951.

[29]   Cavanagh, J., Fairbrother, W.M., Palmer, A.G. and Skelton, N.J. Protein NMR Spectroscopy: Principles and Practice, Academic Press, 1996.

[30]   Baldus, M., Petkova, A.T., Herzfeld, J. and Griffin, R.G.) *Mol. Phys.* 95, (1998) 1197.

[31]   Bloembergen, N. *Physica*, 15 (1949) 386-426.

[32]   Howhy, M., Rienstra, C.M., Jaroniec, C.P. and Griffin, R.G. J. Chem. Phys., 110 (1999) 7983.

[33]   Kubo, A. and McDowell, C.A. J. Chem. Soc., Faraday Trans., 84 (1988) 3713.

[34]   Marassi, F.M., Ma, C., Gratkowski,H., Straus, S.K., Strebel,K., Oblatt-Montal, M., Montal, M., and Opella, S.J. Proc. Nat. Acad. Sci, 96 (1999) 14336.

Determination of Torsion Angles in Membrane Proteins

J. C. Lansing[a], M. Hohwy[a], C. P. Jaroniec[a], A. Creemers[b], J. Lugtenburg[b], J. Herzfeld[c], and R.G. Griffin[a]

[a]Dept. of Chemistry and Francis Bitter Magnet Laboratory
Massachusetts Institute of Technology, Cambridge, MA 02139-4307 USA
[b]Dept. of Chemistry, Leiden University, Leiden, The Netherlands
[c]Dept. of Chemistry, Brandeis University, Waltham MA 02454-9110 USA

Introduction

Bacteriorhodopsin (bR) harnesses light energy to transport protons across the cell membrane of *H. salinarium*. Absorption of a photon by the protonated retinal chromophore initiates a cycle in which the chromophore releases a proton to an aspartate on the extracellular side and reprotonates from an aspartic acid on the cytoplasmic side. Vectorial proton transport depends on a switch in accessibility of the chromophore Schiff base nitrogen from the extracellular to cytoplasmic side. Changes in the retinal conformation are expected to be particularly important for understanding the pumping mechanism. Twists about the chromophore polyene chain (Fig. 1) affect the proton affinity of the Schiff base nitrogen[1] as well as the relative orientation and

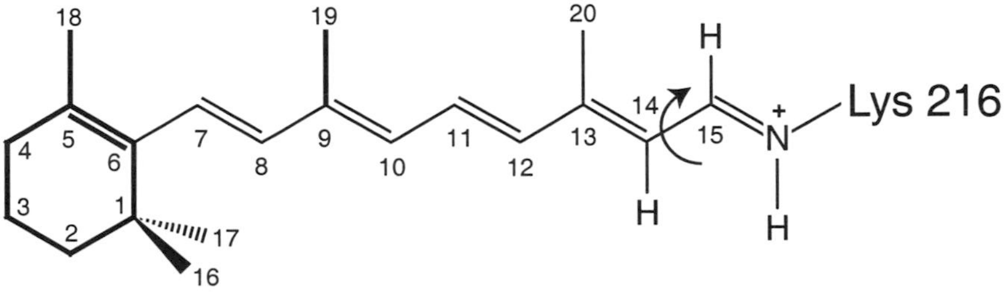

Figure 1: The protonated retinal Schiff base chromophore of bacteriorhodopsin. The measured H-C14-C15-H torsion angle, ϕ, is indicated by the curved arrow.

S.R.Kiihne and H.J.M.deGroot (eds.), Perspectives on Solid State NMR in Biology, 185-190.

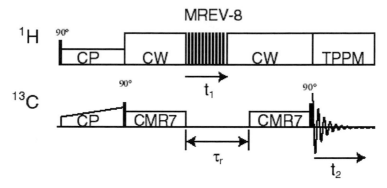

Figure 2: The HCCH pulse sequence, modified from the scheme of Feng et al[11] to incorporate the improved double quantum mixing sequence CMR7.[12] During the constant time of one rotor period, the coherence is allowed to evolve for a time t_1 under the heteronuclear ^1H-^{13}C couplings that are reintroduced by MREV-8[13,14] decoupling. CW irradiation is applied for the remainder of the constant time period to remove heteronuclear couplings.

proximity of proton donor and acceptor groups. In this paper we briefly describe an experiment to measure the twist about various bonds in retinal bound to bR.

FTIR[2] and CD[3,4] measurements indicate distortion in the resting state of the protein, bR_{568}. Intense hydrogen out of plane bending modes indicate this distortion increases during the photocycle in the vicinity of the retinal C15.[2,5-7] Unfortunately, these methods do not provide precise measurements of the chromophore conformation. In contrast, X-ray structures[8-10] of bR_{568} suggest a planar conformation of the C14-C15 bond, with a dihedral angle of 174-180°. To resolve this ambiguity, we utilized solid-state NMR methods to measure the torsion about the C14-C15 bond.

Methods

Solid-state NMR methods permit quantitative measurements of torsion angles through correlation of anisotropic dipolar interactions. Measurement of the H-C-C-H dihedral angle can be accomplished with the experiment shown in Fig. 2.

Evolution of the double quantum signal (sum intensity) in the indirect dimension can be described by

$$a(t_1) = \left\langle \sin^2(\omega_{2Q}\tau_{2Q}) \prod_\lambda \cos \Psi_\lambda(t_1) \right\rangle \tag{1}$$

where ω_{2Q} is the time- and orientation-dependent double quantum nutation rate from zeroth order average Hamiltonian theory,[12] τ_{2Q} is the double quantum excitation time, the index λ denotes a particular ^1H-^{13}C coupling, and the dephasing angle $\Psi_\lambda(t_1)$ is defined by the time average of the dipolar coupling,

$$\Psi_\lambda(t_1) = -\kappa\delta_\lambda \sum_{m=-2}^{2} D_{0,m}^{(2)}\left(\Omega_{PR}^\lambda\right)d_{m,0}^{(2)}\left(\beta_{RL}\right)\int_0^{t_1} e^{-im\omega_r t}dt, \tag{2}$$

where the solid angle Ω_{PR} defines the rotation from the principal axis system of the dipolar tensor to the rotor-fixed frame, β_{RL} is the magic angle, $D_{0,m}^{(2)}$ is a Wigner rotation matrix, $d_{m,0}^{(2)}$ is a reduced Wigner rotation matrix, δ_λ is the magnitude of the static ^{1}H-^{13}C dipolar coupling, and κ is the scaling factor of the MREV-8 sequence. Isotropic chemical shift evolution is neglected by this treatment as placement of the carrier between the ^{13}C resonances results in the evolution of the DQ coherence under a sum chemical shift of zero. Chemical shift anisotropy terms commute with the heteronuclear couplings and are refocused over the constant time of one rotor period, τ_r.

Simulations: Information about the relative orientation (torsion and bond angles) between two couplings is contained in the difference between solid angles $\Omega_{PR}^a - \Omega_{PR}^b$ and can be extracted from experimental data through simulations utilizing the analytical expression in Eq. 1. Signal decay in t_1 due to insufficient ^{1}H-^{1}H decoupling during the MREV-8 sequence can be approximated as exponential decay when $T_2 \gg \tau_r$. Bond angles were taken from the crystal structure of all-*trans* retinal. The value of the scaled heteronuclear dipolar coupling was set to the experimental value of 12.3 kHz measured in a DIPSHIFT[15] measurement of [2-^{13}C]leucine. The influence of the two-bond coupling to the Schiff base proton, the nearest neighbor to the H-C14-C15-H system, was found to be negligible for conformations of the C14-C15 and C15-N bonds that are close to *trans*. Protons even more distant are not expected to influence the signal evolution. Powder averaging was accomplished with 256 pairs of alpha and beta crystallite angles using the REPULSION method.[16]

Sample preparation: Synthesis of [14,15-^{13}C$_2$]retinal, enriched to 99%, was performed according to procedures described elsewhere.[17] Bacteriorhodopsin-containing purple membrane (PM) fragments were isolated from *H. salinarium* according to conventional procedures.[18] The labeled retinal was incorporated by bleaching the sample in 0.5 M hydroxylamine and subsequent regeneration of the apoprotein with the labeled retinal as described elsewhere.[19] Regenerated PM fragments were washed multiple times in 0.3 M guanidine at pH 10.0. Centrifugation of the suspended PM fragments produced a pellet that was packed into a 5 mm quartz rotor. The bR$_{568}$ state was prepared in the custom-designed probe by illumination of the sample with the full visible spectrum of a 1000 W Xenon lamp at 0 °C for 2 hours.

Results and Discussion

The ^{13}C spectrum of the resting state, bR$_{568}$, of [14,15-^{13}C$_2$] retinal-labeled bR is presented in Fig. 3. Evolution of the double-quantum signal intensity (Fig. 4) is

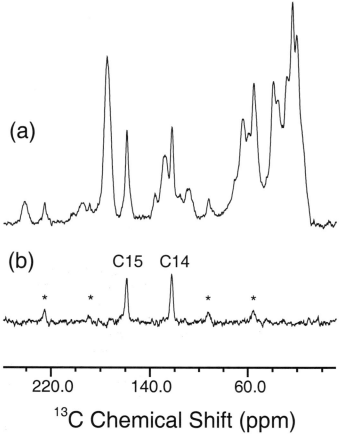

Figure 3: (a) Cross polarization and (b) double-quantum filtered ^{13}C spectra of light-adapted [14,15-$^{13}C_2$]retinal-labeled bR. Rotational sidebands are designated with asterisks. Notice the excellent suppression of the natural abundance background in the spectrum retaining only the coupled spins in the retinal with c.a. 50% efficiency. Spectra were collected on custom-designed spectrometer operating at a ^1H frequency of 317 MHz. The sample was maintained at –90 °C with a sample rotation rate of 5315±2 Hz. Proton fields of 50 kHz for cross polarization, 100 kHz for TPPM decoupling and 127.6 kHz for MREV-8 and CW decoupling were utilized. Carbon fields of 44.7 kHz for cross polarization and 37.2 kHz for CMR7 were applied. The durations of the cross polarization and CMR7 mixing periods were 2 ms and 752.64 ms, respectively. Each t_1 point is the average of 29,696 transients with a recycle delay of 2 s.

indicative of a torsion angle of ±164°. Reduced chi-square analysis indicates an error of ±4° at the 90% confidence level.

These results indicate considerable distortion of the bound chromophore from the planar conformation preferred by free retinal[20] and a protonated retinal Schiff base model compound.[21] That the C14-C15 bond should be twisted from the relaxed

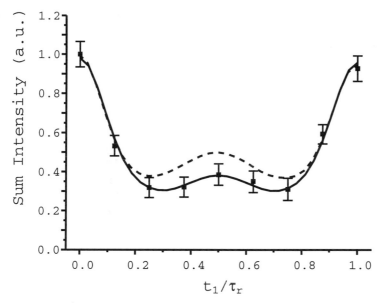

Figure 4: *Modulation of the sum polarization of light-adapted [14,15-$^{13}C_2$] retinal-labeled bR in the indirect dimension of the HCCH experiment. The solid line denotes the best fit to ±164°. The simulated curve for a planar trans conformation (dotted line) is shown for comparison.*

planar structure in the bR resting state by interactions with the binding pocket is surprising. Although nominally a single bond, *ab initio* calculations[1] indicate that the protonated Schiff base C14-C15 bond has a larger barrier of rotation than the neighboring nominal double bonds due to the influence of the conjugated π system. Clearly, it is not safe to assume that ligated retinal will behave in the same fashion as it would in the absence of the protein environment.

Conclusions

Knowledge of the conformation of the retinal in the ground and photointermediate states of bR is crucial to understanding the proton translocation mechanism of this membrane protein. We find that the C14-C15 bond is non-planar, with $|\phi|= 164 \pm 4°$. These results indicate the presence of local structural features unobserved in X-ray structures that have the potential to influence the protein function. More extensive characterization of the chromophore conformation in the bR resting state and upon photoexcitation is warranted to produce a detailed picture of the active site.

We have outlined an experiment based on double quantum excitation and reconversion to measure the ϕ angle about the H-C14-C15-H bond. The experiment cleanly separates the spectrum of the coupled spins from the natural abundance background. It should be equally applicable to other positions in the retinal molecule as well as to measurements of sidechain torsion angles in the protein itself.

Acknowledgements

We thank C.M. Rienstra and B.A. Tounge for stimulating discussions, and D.R. Ruben, A. Thakkar and P. Allen for technical assistance in performing these experiments. The research was supported by grants from the National Institutes of Health (GM-23289, GM-36810, and RR-00995).

References

[1]   Tajkhorshid, E.; Paizs, B.; Suhai, S. *J. Phys. Chem. B 103* (1999) 4518-4527.
[2]   Fahmy, K.; Siebert, F.; Grosjean, M. F.; Tavan, P. *J. Mol. Struct. 214* (1989) 257-288.
[3]   El-Sayed, M. A.; Lin, C. T.; Mason, W. R. *Proc. Natl. Acad. Sci. U. S. A. 86* (1989) 5376-5379.
[4]   Wu, S.; El-Sayed, M. A. *Biophys. J. 60* (1991) 190-197.
[5]   Maeda, A.; Sasaki, J.; Pfefferle, J.-M.; Shichida, Y.; Yoshizawa, T. *Photochem. Photobiol. 54* (1991) 911-921.
[6]   Doig, S. J.; Reid, P. J.; Mathies, R. A. *J. Phys. Chem. 95* (1991) 6372-6379.
[7]   Pfefferle, J.-M.; Maeda, A.; Sasaki, J.; Yoshizawa, T. *Biochem. 30* (1991) 6548-6556.
[8]   Luecke, H.; Schobert, B.; Richter, H.-T.; Cartailler, J.-P.; Lanyi, J. K. *J. Mol. Biol. 291* (1999) 899-911.
[9]   Essen, L.-O.; Siegert, R.; Lehmann, W. D.; Oesterhelt, D. *Proc. Natl. Acad. Sci. U. S. A. 95* (1998) 11673-11678.
[10]   Sass, H. J.; Buldt, G.; Gessenich, R.; Hehn, D.; Neff, D.; Schlesinger, R.; Berendzen, J.; Ormos, P. *Nature 406* (2000) 649-653.
[11]   Feng, X.; Lee, Y. K.; Sandstrom, D.; Eden, M.; Maisel, H.; Sebald, A.; Levitt, M. H. *Chem. Phys. Lett. 257* (1996) 314-320.
[12]   Rienstra, C. M.; Hatcher, M. E.; Mueller, L. J.; Sun, B.; Fesik, S. W.; Griffin, R. G. *J. Am. Chem. Soc. 120* (1998) 10602-10612.
[13]   Rhim, W.-K.; Elleman, D. D.; Vaughan, R. W. *J. Chem. Phys. 59* (1973) 3740-3749.
[14]   Mansfield, P. *J. Phys. C 4* (1971) 1444-1452.
[15]   Munowitz, M.; Aue, W. P.; Griffin, R. G. *J. Chem. Phys. 77* (1982) 1686-1689.
[16]   Bak, M.; Nielsen, N. C. *J. Magn. Reson. 125* (1997) 132-139.
[17]   Pardoen, J. A.; Winkel, C.; Mulder, P. P. J.; Lugtenburg, J. *Recl. Trav. Chim. Pays-Bas 103* (1984) 135-141.
[18]   Oesterhelt, D.; Stoeckenius, W. *Methods Enzymol. 31* (1973) 667-678.
[19]   Hu, J. G.; Sun, B. Q.; Bizounok, M.; Hatcher, M. E.; Lansing, J. C.; Raap, J.; Verdegem, P. J. E.; Lugtenburg, J.; Griffin, R. G.; Herzfeld, J. *Biochem. 37* (1998) 8088-8096.
[20]   Hamanaka, T.; Mitsui, T.; Ashida, T.; Kakudo, M. *Acta Crystallogr., Sect. B 28* (1972) 214-222.
[21]   Santarsiero, B. D.; James, M. N. G.; Mahendran, M.; Childs, R. F. *J. Am. Chem. Soc. 112* (1990) 9416-9418.

Characterization and assignment of uniformly labeled NT(8-13) at the agonist binding site of the G-protein coupled neurotensin receptor

P.T.F. Williamson[a,c], S. Bains[b], C.Chung[b], R.Cooke[b], B.H.Meier[a], A.Watts[c]

[a]Laboratorium für Physikalische Chemie, Universitatstrasse 22, ETH-Zentrum, CH-8092 Zürich, Switzerland. Tel No. +41-1-632-4404 Fax No. +41-1-632-1021.
[b]Glaxo-Wellcome, Medicines Research Centre, Gunnels Wood Road, Stevenage, Hertfordshire.SG1 2NY UK
[c]Biomembrane Structure Unit, Biochemistry Dept., University of Oxford, South Parks Road, Oxford, OX1 3QU, UK

Introduction

The neurotensin receptor is a member of the G-protein coupled receptor (GPCR) family of transmembrane proteins that is activated upon the binding of the basic tridecapeptide agonist, neurotensin, to the extracellular surfaces of cells. The neurotensin receptor is found widely in both the central nervous system and the periphery. In the periphery, it stimulates smooth muscle contraction [1,2]. In the central nervous system, it mediates a variety of activities including antinociception, hypothermia and increased locomoter activity [3-5]. These effects are probably mediated through the regulation of the mesolimbic and negrostriatal dopamine pathways [6,7]. As a result, the pharmacological action of the neurotensin is similar to that observed for dopamine, where compounds function as antipsychotics [6,7], and intervention may provide useful insights for the development of treatments for conditions such as schizophrenia [8] and Parkinson's disease [9].

To date, no direct high resolution structural information is available for the neurotensin receptor due to limited successes at production of 2D and 3D crystals for diffraction studies and to the unfavorable relaxation rates associated with this size of membrane system in conventional high resolution solution state NMR studies. Although no direct structural information is available for the neurotensin receptors, holistic modeling approaches have been employed to provide evidence that the receptor

S.R.Kiihne and H.J.M.deGroot (eds.), Perspectives on Solid State NMR in Biology, 191-201.
©2001 Kluwer Academic Publishers. Printed in the Netherlands.

adopts the typical 7 transmembrane motif shared by this family of receptors [10]. A putative agonist binding site containing the agonist analogue neurotensin(8-13) has been modeled, supported by site directed mutagenesis and structure/activity studies [11]. High-resolution solution NMR structural studies of the agonist, neurotensin, in the absence of receptor have revealed that no preferred conformation exists in solution [12]. Extensions of these studies to neurotensin in the presence of the membrane mimetic sodium dodecyl d_{25}-sulphate again indicate that no preferred conformation was adopted although some ordering of charged residues on the surface of the micelles was observed [12].

The aim of the studies presented here is to provide direct structural information for the agonist, neurotensin, whilst bound to the functionally active neurotensin receptor. This information may be included in future models to aid the understanding of the events associated with the molecular recognition of neurotensin by the neurotensin receptor. To this end we have undertaken the preparation of sufficient quantities of recombinant neurotensin receptor to support a solid state NMR study. Through the incorporation of both carbon-13 and nitrogen-15 into a pharmacologically active C-terminal fragment of neurotensin, neurotensin(8-13) (Fig 1), we have been able to employ solid state NMR methodologies to specifically observe neurotensin(8-13) whilst bound to the neurotensin receptor and to assign resonances to particular groups within the agonist. As a prelude to a full assignment of neurotensin(8-13) bound to the receptor, we have performed a near complete assignment of the lyophilized agonist fragment. This demonstrates that sufficient resolution exists to permit a full assignment of such compounds. These assignments are essential to further structural work and may, under favorable conditions, allow us to suggest a preferred conformation of the agonist whilst bound to the receptor.

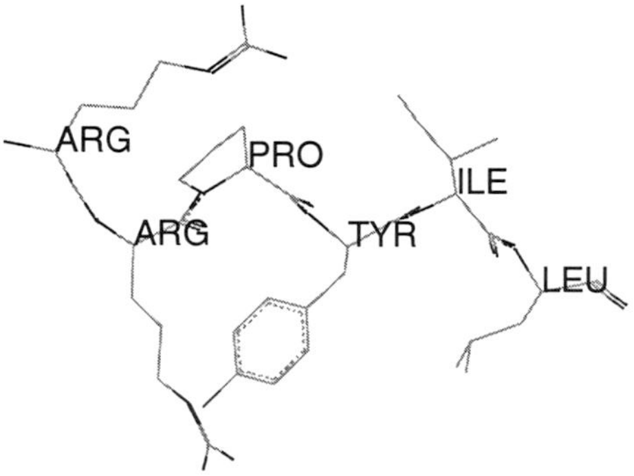

Fig 1. Diagram showing an energy minimized structure of neurotensin(8-13) (Arg[1]-Arg[2]-Pro[3]-Tyr[4]-Ile[5]-Leu[6]), a C-terminal fragment of neurotensin that shows similar pharmacological efficacy (Kd=13nM).

Materials and Methods

Expression of neurotensin receptor. Detergent solubilized rat neurotensin receptor was obtained by the method of Grisshammer [13]. For expression *E.coli* strain DH5α was grown on double strength TY medium [14] containing ampicillin (100 µg/ml) and 0.2% glucose. A NTR fusion protein was expressed from the pRG/III-hs-MBPP-T43NTR-TrxA-H10 gene construct kindly donated by Dr R. Grisshammer [13]. This plasmid contains a truncated neurotensin receptor, which is fused at its N-terminus to a maltose binding protein and its signal sequence (MBP) in order to target the N terminus to the periplasm. It is also fused at its C-terminus to thioredoxin, to aid stability, and to a deca-his tag to assist in purification. The transformed DH5α were grown in 400ml of medium in a 1 l flask at 37°C until OD660 reached 0.7. The cultures were subsequently induced with 0.5mM isopropyl-β-galactoside (IPTG). The temperature was then lowered to 20°C and incubated for a further 40 h. The cells were harvested by centrifugation, flash frozen in liquid nitrogen and stored at –70°C.

Purification of neurotensin receptor. The purification of heterologously expressed neurotensin receptor was performed using the method of Grisshammer [15]. Briefly, 200g of cell paste were resuspended in 1.2l of neurotensin buffer (50mM TRIS, 0.2M NaCl, 30% glycerol, 0.5% CHAPS, 0.1% CHS, 0.1% LM) containing pepstatin A, leupeptin A, PMSF, lysozyme and Dnase, to prevent protein degradation and to aid cell lysis. The cells were then broken by three passes through a flow-through sonicator and subsequently clarified by centrifugation. The supernatent was loaded onto a Quiagen NTA nickel affinity column in 1mM imidazole at a flow rate of 10 ml min⁻¹. The neurotensin receptor eluted with neurotensin buffer containing 350 mM imidazole. The eluate was then concentrated using an Amicon stirred cell with YM-30 membrane.

The buffer was subsequently exchanged to a low salt buffer (50mM TRIS, 20mM NaCl, 30% glycerol, 0.5% CHAPS, 0.1% CHS, 0.1% LM) using a 150 ml Sephadex-G25 column. The active receptor was purified using a neurotensin affinity column[13]. The affinity column was equilibrated with low salt buffer and the fraction containing the neurotensin receptor loaded at 0.5 ml min⁻¹. Following washing with both low salt buffer and 200 mM KCl buffer, the active neurotensin receptor was eluted using a high salt buffer (50 mM TRIS, 1.0 M NaCl, 30% glycerol, 0.5% CHAPS, 0.1% CHS, 0.1% LM). Prior to NMR studies the purified neurotensin receptor was returned to desalting buffer and concentrated using a stirred cell Amicon and Centricon containing a YM-30 membrane.

The activity of the protein was monitored using a tritiated-neurotensin binding assay [15], whilst the protein concentration was monitored using an amido black protein assay [15]. Purity was determined by a 5-12% gradient SDS-PAGE, followed by Comassie staining [15].

Solid phase synthesis of neurotensin(8-13). Neurotensin(8-13) was synthesized using conventional FMOC solid phase synthesis at the Oxford Center for Molecular Sciences.

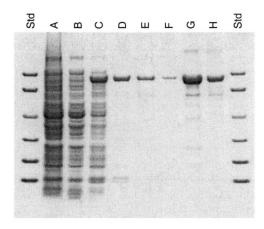

Fig 2. 4-12% Comassie stained MES-SDS polyacrylamide gradient gel of samples obtained during the purification of recombinant neurotensin receptor. lanes: standards (Std), clarified sample(A), Quiagen-NTA flow through (B), 500mM imidazole eluate (C), desalted sample (D), NT column flow through (E), 200mM KCl wash (F), 1M NaCl eluate from NT affinity column (G) and concentrated sample of neurotensin receptor (H).

Uniformly ^{13}C and ^{15}N labeled amino acids (Promochem, UK) were protected and purified using standard amino acid protection protocols [16,17]. The compounds confirmed by electrospray mass spectroscopy and thin layer chromatography [17]. Following solid phase synthesis, the peptide was purified by reverse phase HPLC, eluting at an acetonitrile concentration of 27% comparable with unlabeled neurotensin(8-13) (Sigma, UK) used as a standard. Electrospray mass spectroscopy of the final product gave a single molecular species with molecular weight 868 Da consistent with uniform ^{13}C and ^{15}N isotopic labeling of neurotensin(8-13).

NMR methods. CP-MAS spectra of the lyophilized samples of neurotensin(8-13) were acquired on a Bruker Avance 600, operating at 600 MHz (proton Lamour frequency) with a 2.5mm Bruker triple resonance probehead. Carbon-13 and Nitrogen-15 cross polarization magic angle spinning (CP-MAS) spectra were acquired using an adiabatic cross polarization sequence [18] with a constant proton field of 60 kHz. Decoupling during acquisition was performed with 125 kHz TPPM decoupling [19] (phase alternation 7°). Carbon-13/Nitrogen-15 HETCOR experiments was performed at 15 kHz spinning, and transfer from Nitrogen-15 and Carbon-13 was achieved by an adiabatic sweep (10ms, centered at 50 kHz field). During t_1 and t_2, TPPM decoupling was applied as described above, and Lee Goldburg decoupling with an applied field of 120kHz was employed during mixing. Phase sensitive detection in t_1 was achieved using TPPI phase cycling of the initial proton to nitrogen-15 cross polarization. Carbon-13/Carbon-13 correlation data was acquired using a 2D exchange experiment at 22.5 kHz spinning with a RFDR sequence [20] (π pulses of 8µs were applied rotor

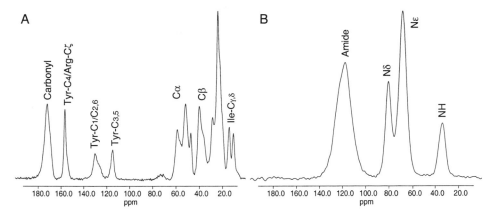

Fig 3. Carbon-13 (A) and nitrogen-15 (B) CP-MAS spectra of neurotensin(8-13) uniformly labeled with carbon-13 and nitrogen-15. Both spectra acquired with CP-MAS at 15 kHz spinning speed with other parameters as described in the text. Data accumulated over 128 and 256 acquisition respectively and processed with 30 Hz linebroadening.

synchronously with XY-8 phase cycling) applied during the mixing period. During t_1 and t_2, TPPM decoupling was applied as described, and 120 kHz Lee-Goldburg decoupling was applied during mixing.

Carbon-13 CP-MAS spectra of detergent solubilized receptor were obtained on a Chemagnetics CMX-500 operating at 500 MHz (proton Larmour frequency), with a Chemagnetics 6mm triple resonance probehead. CP-MAS spectra were acquired using an adiabatic cross polarization sequence with a proton field of 60 kHz. During acquisition protons were decoupled using 80 kHz continuous wave irradiation. Double quantum-filtered experiments were performed using the POST-C7 sequence [21]. The carbon-13 B_1 field (35kHz) was matched to seven times the rotor speed and 10 C7 elements were applied for both excitation and reconversion. Double quantum coherence is selectively observed through the appropriate phase cycling of the C7 reconversion sequence, the final $\pi/2$ pulse and the receiver [21]. During the C7 sequence the protons were decoupled using Lee-Goldburg decoupling with an applied field of 80 kHz.

Results & Discussion

Purification of detergent solubilized neurotensin receptor. Recombinant neurotensin receptor was purified from 230 l of culture (~1.4kg wet cell paste) in seven batches. Typically crude cell lysate contained neurotensin receptor at 1-3 pmoles mg^{-1}. A 200 fold enrichment in the neurotensin receptor (~600 pmoles mg^{-1}) was achieved by Ni affinity purification, which removed the bulk of the contaminating proteins. The remaining contaminants were non-function neurotensin receptor and endogenous E.coli Ni affinity binding proteins. These were removed by purification on a neurotensin affinity column, which resulted in a further 3.5 fold enrichment of the neurotensin receptor (specific activity 2500 pmoles mg^{-1}). The course of the purification and final purity of the recombinant neurotensin receptor was monitored by SDS-PAGE and is shown in Figure 2 (Final sample, Lane H).

Assignment of neurotensin(8-13). The carbon-13 and nitrogen-15 CP-MAS spectra of neurotensin(8-13) are shown in Figures 3A and 3B, respectively. Although in the 1D CP-MAS spectra, individual sites are rarely resolved due to the large linewidths observed in this heterogeneous sample of lyophilized neurotensin(8-13), these broad resonances can be assigned to the individual functional groups which contribute to the neurotensin(8-13) (Fig 3A and 3B). In the carbon spectrum, several sites can be assigned on the basis of their unique chemical shifts including the tyrosine, arginine and some of the aliphatic side chains. However many of the resonances associated with the

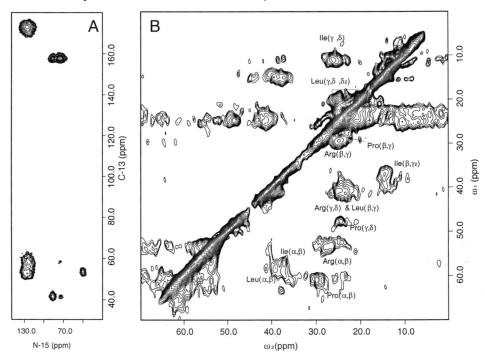

Fig 4. _Nitrogen-15/Carbon-13 heteronuclear correlation experiment of lyophilized neurotensin(8-13)(A). Data acquired with 128 t_1 points with 64 acquisitions for each point. Data processed with 50 Hz linebroadening prior to Fourier transform. Data in t_1 was then linear predicted from 128 to 512 points and a sinebell function applied prior to a real Fourier transform in t_1. Final matrix 2048x2048 points. Expansion of the sidechain region of a ^{13}C-^{13}C-correlation spectrum (B) of lyophilized neurotensin(8-13) with 2.3 ms exchange at 22.5 kHz. Data acquired with 256 t_1 points and 128 acquisitions of each. Data processed with 30 Hz linebroadening in t_2. Data was linear predicted from 256 to 1024 points and processed with a sinebell function prior to Fourier transformation in t_1. Final matrix 2048x2048 points._

Table 1. Carbon-13 assignments for lyophilized neurotensin(8-13) obtained from both homo-and heteronuclear correlation experiments.

Residue	CO	C_α	C_β	C_γ	C_δ	C_ε	C_ζ	Other
			Carbon-13 chemical shift ($\pm$1.0ppm)					
Arg_1	169.6	53.3	28.9	24.5	40.9		157.6	
Arg_2		52.6	28.9	24.5	40.9		157.6	
Pro_3	172.3	61.1	29.5	24.1	48.0			
Tyr_4	174.2	52.8	N/A					127.9 (C1)
								130.7(C2,6)
								115.7(C3,5)
								155.6(C4)
Ile_5	172.3	57.4	36.8	25.6	11.0			
				14.2				
Leu_6	172.5	53.4	38.7	23.7	18-27			
					18-27			

peptide backbone are poorly resolved due to the relatively small chemical shift dispersion and the relatively large linewidths associated with the sample. A similar situation is apparent in the nitrogen-15 spectrum where the arginine sidechains and terminal amine are clearly visible while poor resolution is apparent in the amide region centered at 125 ppm. To circumvent these problems of poor resolution and to provide essential correlation data to allow the assignment of these 1D spectra, both homo- and heteronuclear correlation spectra were acquired.

The nitrogen-15/carbon-13 correlation spectrum is shown in Figure 4A. The spectrum is dominated by the strong correlations between the amide nitrogen (120ppm) and the carbonyl (175ppm) and C_α resonances (~52ppm). In these regions, areas of heightened intensity can be seen that correspond to the C_α/CO correlations observed in the homonuclear correlation spectra (See below). In addition, strong correlations are also observed between the N-terminal amine (~37ppm) and the C_α of Arg_1 (53.3ppm). Strong correlations between the two inequivalent nitrogens of the arginine sidechain, N_ε (~84ppm) and N_δ (~70ppm), and the C_ζ carbon are also apparent. The N_δ can then be traced to C_δ(40.9ppm). Under the long mixing conditions employed here, it is also apparent that long range transfer occurs between the N_ε and the C_δ. Whether this is due to relayed transfer along the chain or to direct through space transfer is unclear.

An expansion of the sidechain region (0-70ppm) of a carbon-13/carbon-13 correlation spectrum of neurotensin(8-13) is shown in Figure 4B. Under the regime chosen with short exchange times, the intensities observed arise almost solely from correlations between direct carbon-carbon pairs. Several regions of intensity in the spectrum arise primarily from the C_α/CO connectivities: the tyrosine ring region and the upfield region assigned to the C_α and aliphatic sidechains. On the basis of the unique chemical shifts arising from several of the sites in the sidechains, it has been possible to assign most of the sidechain resonances and to correlate them to the backbone C_α and

CO resonances, allowing an almost complete assignment of the lyophilized neurotensin(8-13). The results of this assignment are shown in Table 1. With the exception of Tyr_β and the Leu_δ, it has been possible to assign all sites unambiguously. We attribute our inability to assign the two Leu C_δ to the relatively small chemical shift dispersion expected between the two resonances which prohibits their resolution against the more intense diagonal elements in the spectrum. Hightened intensity in this region does indicate that these sites are correlated (Figure 4A, box). In contrast, the absence of the $Tyr\text{-}C_\beta$ assignment arises due to the absence of any intenisty in the $Tyr\text{-}C_\beta/Tyr\text{-}C1$ region possibly due to unfavourable relaxation under the mixing sequence chosen.

Assignment of neurotensin(8-13) bound to detergent solubilized neurotensin receptor.
The carbon-13 CP-MAS spectrum of detergent solubilized neurotensin receptor is shown in Fig 5A. The spectrum is dominated by the strong natural abundance signals arising from the detergents (170-180ppm, ~100ppm and 10-50ppm) and glycerol (62 and 71 ppm) present in the sample. Upon the addition of a stoichiometric amount of neurotensin(8-13) (10 nmoles, <5% free ligand), the major spectral features remain constant (Fig 5B). In regions displaying greater spectral complexity (0-70 ppm), we are unable to identify any additional resonances which may arise from the C_α, C_β, and C_γ resonances of the ligand as they are masked by the natural abundance signal from the buffer components. In regions that are less crowded, such as between 169 and 175 ppm and 152 and 157 ppm, spectral perturbations are observed. The spectral intensity arising between 169 and 175ppm has been assigned to the labeled carbonyl groups present in the peptide, whilst the spectral intensity between 152 and 157ppm has been assigned to the C_ξ of the arginine sidechains and the C4 of the tyrosine sidechains on the basis of the observed chemical shifts. CP-MAS studies of NT(8-13) in identical buffers at these temperatures revealed no such intensities suggesting that the signal arising in these spectra occur due to the additional immobilization inferred to the ligand upon binding to the detergent solubilized receptor. Additionally higher resolution spectra acquired with direct carbon excitation and MAS at 5°C indicated significant perturbations in chemical shift upon the binding of the neurotensin(8-13) to the receptor [17] (data not shown). From these studies we conclude that the resonances observed arise solely from the neurotensin(8-13) specifically bound to the neurotensin receptor.

To resolve sites masked by the background natural abundance signal from the detergents, double quantum filtered spectra of the sample were obtained (Fig 5C) through the reintroduction of dipolar couplings using the POST-C7 sequence. Through the specific observation of only those nuclei that have strong carbon-carbon couplings, it has been possible to suppress the large natural abundance background signal sufficiently (as only 0.01% of natural abundance carbon atoms share labeled neighbors). This has allowed us to observe resonances arising primarily from the C_α(50-60ppm) and from sites in the aliphatic sidechains (10-50ppm) within the peptide in addition to those previously observed for both the aromatic and basic sidechains. The linewidths

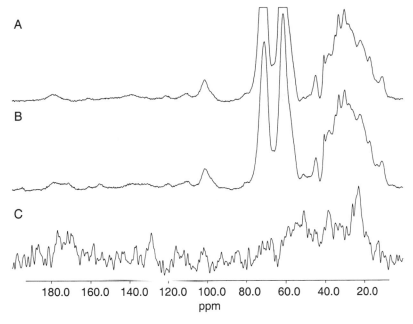

Fig 5. Carbon-13 CP-MAS spectra of 1 mg of detergent solubilized neurotensin receptor (A) and upon the addition of a stoichiometric amount of neurotensin(8-13) (10 nmoles) (B). Data acquired with a 1ms contact time, 5000 Hz spinning speed and 70 kHz decoupling. POST-C7 double quantum filtered spectra of 1 mg of detergent solubilized neurotensin receptor containing 10 nmoles of uniformly labeled neurotensin(8-13), acquired with a 512 μs excitation and reconversion period. Data averaged over 8192 acquisitions and processed with 30 Hz linebroadening (C - 100Hz linebroadening).

observed for the bound neurotensin(8-13) in the 1D CP-MAS experiments are similar to those observed for neurotensin(8-13). As we have demonstrated for the lyophilized sample above, these conditions are sufficient to allow a near complete assignment of the neurotensin(8-13) whilst bound to the neurotensin receptor under solid state conditions.

Conclusions

The data presented here demonstrate that we are able to suppress large natural abundance background signals to selectively observe the agonist analogue neurotensin(8-13) whilst resident in the agonist binding site on the detergent solubilized neurotensin receptor. This is the first time an exchangeable ligand has been observed whilst bound to a GPCR by solid state NMR. Even in the absence of a suitably crystalline sample, through the application of homo-nuclear and hetero-nuclear correlation spectroscopy, we have demonstrated that the methodology and the resolution exists to permit a full assignment of the neurotensin(8-13) whilst bound to the receptor under solid state conditions. This data demonstrates that upon the successful reconstitution of the the neurotensin receptor into lipid bilayers the methodology exists to permit a partial assignment of the bound neurotensin(8-13) whilst resident in the agonist binding site of the reconstituted receptor. Prior to further structural studies, we aim to exploit the chemical shifts observed to provide information relating to the conformation of the ligand as a lyophilized powder and whilst bound to the receptor.

Acknowledgements

We wish to acknowledge the invaluable assistance of R. Grisshammer for his help on the expression and purification of the neurotensin receptor. We acknowledge M. Peatkeathly (Oxford Centre for Molecular Sciences) for her synthesis of the neurotensin(8-13). We would like to acknowledge G. Gröbner, P.J.R. Spooner and A. Detken for their useful discussions on the recoupling sequences used. This work was supported by a BBSRC-CASE Glaxo-Wellcome studentship.

References

[1]   Kitagbi P., Ann. NY Acad. Sci. 400 (1982) 37-55
[2]   Kachur, J.F., Miller, R.J., Field, M. and Revier, J., J. Pharmacol. Exp. Ther. 220 (1982) 456-483
[3]   Osbahr, A.H., Nemeroff, P.J., Manberg, P.J. and Prange, A.J., Eur. J. Pharmacol. 54 (1979) 299-302
[4]   Ervin, G.N., Birrema, L.S., Nemeroff C.B. and Prange, A.J., Nature 291 (1981) 73-76
[5]   Osbahr, A.J., 217 (1981) J. Pharmacol. Exp. Ther. 465-651
[6]   Garcia-Sevilla, J.A., Magnusson, T., Carlsson, A., Leban, J. and Folker, K., Arch. Pharmacol. 305 (1978) 213-218
[7]   Widerlov, A., Kilts, C.D., Mailman, R.B., Nemeroff, C.B., Prange, A.J. and Breese, G.R., J. Pharmacol. Exp. Ther. 223 (1982) 1-6
[8]   Reches, A., Burke, R.E, Jiang, D., Wagner, H.R. and Fahn, S., Peptides 4 (1983) 43-48
[9]   Uhl, G.R., Whitehouse, P.J. and Price, W.W., Brain Res. 308 (1984) 186-190
[10]   Vriend, G., J. Mol. Graphics. 272 (1997) 144-164
[11]   Pang, Y.P., J. Biol. Chem. 271 (1996) 15060-15068
[12]   Xu, G.Y., and Deber, C.M., Int. J. Peptide. Res. 37 (1991) 528-535
[13]   Tucker, J. and Grisshammer, R., Biochem. J. 317 (1996) 891-899
[14]   Sambrook, J., Fritish, E.F. and Maniatis T., Molecular Cloning, Cold Spring Harbor Laboratory Press, 2nd Edition (1992)
[15]   Tucker, J. and Grisshammer, R., Protein Expression and Purification 11 (1997) 53-60
[16]   Jones, J., Amino Acid and Peptide Synthesis, Oxford Chemistry Primers, 1st Ed., Oxford University Press: Oxford, 1992.
[17]   Williamson, P.T.F. D.Phil Thesis, Univeristy of Oxford, 1999
[18]   Baldus, M, Geurts, D.G., Hediger, S. and Meier, B.H., J. Mag. Res. 118 (1996) 140-144
[19]   Bennett, A.E., Reinstra, C.M., Auger, M., Lakshmi, K.V. and Griffin R.G., J. Chem. Phys. 103 (1995) 6951-6955
[20]   Sodickson, D.K., Levitt, M.H., Vega S. and Griffin, R.G., J. Chem. Phys 98 (1993) 6742-

6748
[21] Hohwy, M., Jakobsen, H.J., Eden, M., Levitt, M.H. and Nielsen, N.C. 108 (1998) 2686-
      2694

Structural insight into the interaction of amyloid-β peptide with biological membranes by solid state NMR

Gerhard Gröbner[a], **Clemens Glaubitz**[b], **Philip T. F. Williamson**[c], **Timothy Hadingham**[d], **Anthony Watts**[d]

[a] *Biophysical Chemistry Department, Umea University, SE-90187 Umea, Sweden;*
gerhard.grobner@chem.umu.se
[b] *Physical Chemistry Department, University of Stockholm, Sweden.*
[c] *Physical Chemistry Department, ETH Zurich, Switzerland.*
[d] *Biomembrane Structure Unit, University of Oxford, UK.*

Introduction

Alzheimer's disease (AD) is a chronic dementia, affecting an increasingly large number of old people worldwide [1-3]. AD together with mature onset diabetes and prion-transmissible spongiform encephalopathies, belongs to a category of amyloid diseases, which are all categorized by an abnormal folding of a normally soluble protein into neurotoxic aggregated structures [3-5]. The key event in AD is the metabolism of amyloid precursor protein to amyloid-β-peptide (Aβ) and the subsequent deposition of Aβ as plaques in the brains of patients. This 39-42 amino acid peptide has been linked to the apoptosis of neuronal cells, and its neurotoxicity seems to be associated with its ability to convert from a non-toxic monomeric form into toxic aggregates [5-7]. However the cellular mechanism involved in mediating the toxic effect of Aβ peptide remains unclear [6-11]; and also the process of transformation into insoluble, neurotoxic peptide aggregates. Due to the complexity and dependence of this process on physiological parameters, various models for fibril formation are studied at present including aggregation in solution [7,8], lipid-mediated aggregation of Aβ in contact with cell membrane surfaces [10-13], and formation of transmembrane ion channel-like structures in neuronal membranes [9,14,15]. Structural and biophysical studies of the self-assembly of Aβ-peptide into fibrillar structures found this process strongly dependent on the physical conditions [5,6-8,16]. While earlier studies proposed antiparallel-β-sheet structures for the amyloid fibrils [6], more recent work indicates an

S.R.Kiihne and H.J.M.deGroot (eds.), Perspectives on Solid State NMR in Biology, 203-214.

in register, parallel organization of β-sheets propagating and twisting along the fibrillar axis [5,17]. However, there is growing evidence that the toxic agent is not the mature fibrils themselves, but rather their precursor forms called diffusible "protofibrils" [3,4,8].

Various experimental evidence indicates that non-specific interactions of Aβ with cell membranes may play an important role in AD [9-15]. Since Aβ peptide comprises extracellular and transmembrane (28-42 position) domains, its association with membranes has been shown to induce and accelerate formation of prefibrillar and fibrillar structures. It can also insert into lipid bilayers and form cation-selective channels [9,14,15]. The mechanism of self-assembly of Aβ on membrane surfaces or as ion-channels in membranes is not understood yet, primarily due to the lack of structural information at an atomic level. However, to extract this information is very challenging since any structural biology method including NMR has to deal with a very complex, non-cyrstalline, and disordered system.

Here, we report a strategy for structure determination of Aβ in membranes. First, lipid-modulated structural and aggregational features of Aβ and its interactions with membranes were characterized by circular dichroism (CD) and ^{31}P MAS NMR spectroscopy. These results formed then the foundation for the use of rotational resonance (RR) ^{13}C CP MAS NMR for a first insight into the membrane-bound secondary structure of the peptide, before major aggregation occurs [19-22]. In this context also limits and future prospects of solid state NMR methods for structure determination on these systems are discussed.

Methods

Materials: L-α-Dimyristoylphosphatidylcholine (DMPC) and Dimyristoylphosphati-dylglycerol (DMPG) were obtained from Sigma (UK), DMPC-d_{67} from Avanti Polar Lipids (US). Aβ$_{1-40}$ was synthesized by standard solid-phase FMOC chemistry (NSR Centre, Nijmegen, Netherlands), subsequently purified by HPLC and quality checked by MALDI MS. Aβ$_{1-40}$ containing 1-^{13}C-Ile$_{31}$ and 2-^{13}C-Gly$_{33}$ (Promochem, UK) was prepared in the same way. To obtain a monomeric, soluble form of the peptide, 10 mg peptide was dissolved in 500 μl TFA (trifluoroacetic acid). After removal of TFA by nitrogen stream, TFE (trifluoroethanol) was added to resuspend the protein film and then evaporated under fine vacuum to remove any traces of acidic TFA. For binding studies of Aβ to membrane surfaces, peptide was added to vesicles of various DMPC/DMPG compositions to give a final 30:1 P/L molar ratio. The mixture was incubated for 30 min at 310K, three times freeze-cycled and pelleted. For reconstitution trials, incorporation of Aβ into various DMPC/DMPG bilayers at 30:1 L/P molar ratio was carried out as described before [20]. For incorporation of Aβ peptide in a nonaggregated state into membranes at a 20:1 L/P molar ratio for NMR experiments, 15 mg peptide film was dissolved in TFE (2 ml) subsequently mixed with DMPC-d_{67}, dried as a lipid/peptide film and resuspended in buffer (10 mM NaH$_2$PO$_4$, 0.2 mM EDTA,140 mM NaCl, pH 7.8). After sucrose density purification, vesicles were pelleted into MAS NMR rotors and kept frozen prior to measurements.

CD-measurements: Samples were sonicated under cooling using a probe-type sonicator and metal debris removed by centrifugation. CD-spectra (Jasco, USA) were obtained using a 1mm path length quartz cell (Hellma, Germany). CD-spectra were corrected for the lipid vesicle background and analyzed using the k2d software [23].

NMR experiments: ^{31}P MAS NMR experiments were carried out under efficient proton decoupling (30 kHz), at 81 MHz phosphorous frequency on a 200 MHz Infinity (Chemagnetics, USA) using a double resonance 7 mm MAS NMR probe (Bruker, D). ^{13}C MAS NMR experiments were performed at 100.6 MHz and 125.7 MHz ^{13}C frequencies on Bruker and Chemagnetics spectrometers using double resonance 7 mm and 4 mm MAS Probes. Cross polarization (CP) contact time was 0.6 ms for solid and 1.0 ms for membrane samples. Decoupling power varied between 60-80 kHz. The C_α-glycine ^{13}C resonance was selectively inverted using a DANTE pulse sequence [20], followed by a variable mixing time (0.5 ms – 30 ms).

Results and Discussions

The secondary structural features and aggregation properties of Aβ-peptide are very sensitive to interactions between the peptide and its membrane-environment. CD- and ^{31}P MAS NMR experiments were carried out using lipid vesicles of various compositions to study the structural changes in the peptide as a function of its interactions with membranes either by contact to the surface or by incorporation. In this way suitable starting conditions were found to perform first ^{13}C RR CP MAS NMR experiments to explore structural features in the transmembrane part of the Aβ$_{1-40}$ peptide before major aggregation occurs.

CD-Measurements: How the different lipid-peptide interactions affect the structure of Aβ-peptide can be seen in Figure 1 where results are shown for CD-measurements carried out on Aβ$_{1-40}$ either bound to various membrane surfaces or incorporated into them under different conditions. In Figure 1 (left), CD spectra are displayed for the spectral region between 200 – 240 nm at RT. Trace a) reveals a typical α-helical structure of Aβ, prepared as a monomer in TFE after HPLC purification. Trace b) reveals a significant amount of β-structures for Aβ added to charged membrane surfaces (33mol% negatively charged DMPG) at a 30:1 P/L molar ratio. A similar situation is seen in Trace c) when Aβ was incorporated at the same ratio into the same membrane matrix by reconstitution via dialysis. To the contrary, reconstitution of the peptide into DMPC bilayers by cosolubilization using the membrane mimicking solvent TFE showed predominantly helical features (Trace d). These results are not surprising since TFE stabilizes helical structures while in aqueous conditions a conversion from an

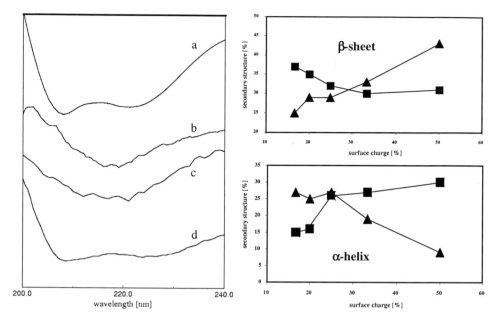

Figure 1. Left: CD-spectra for Aβ₁₋₄₀ peptide at RT (relative mean residue ellipticities scaled and shifted to for better comparison); a) Aβ in TFE; b) Aβ bound to mixed DMPC/DMPG (2:1 PC/PG ratio) membrane surfaces at 30:1 L/P molar ratio; c) Aβ incorporated into mixed DMPC/DMPG (2:1 PC/PG ratio) at 30:1 L/P molar ratio by dialysis reconstitution; d) Aβ incorporated into DMPC membranes by cosolubilization at 20:1 lipid/peptide molar ratio. Right: Lipid-induced fraction of β-sheet and α-helix structures for Aβ added to (▲) or incorporated into (■) mixed bilayers at 30:1 L/P molar ratio. Membrane surface charge varied between 50% and 17%.

initially random coil form into a β-sheet state can easily take place [10]. Interactions of Aβ in aqueous environment with charged membrane surfaces accelerate this conversion as described in detail by Terzi et al. [10] and others. The spectrum obtained for Aβ upon incorporation via dialysis (Trace c) is therefore not unexpected. Reconstitution trials using neutral DMPC alone failed to incorporate Aβ-peptide, probably due to the lack of stabilization of the positively charged peptide in the aqueous micellar detergent system by negatively charged lipid headgroups.

    To study in more detail the relationship between the structural properties of Aβ and the relevant lipid-peptide interactions, comparative binding and incorporation studies were performed under a systematic variation of the lipid environment. The amount of negatively charged DMPG lipid in the DMPC bilayer was varied between 17 mol% and 50 mol%. The analyis of the CD-studies of Aβ either added to or incorporated into these membranes are displayed in Figure 1 (right). Adding Aβ₁₋₄₀ to mixed liposomes, reveals a strong relationship between the amount of β-sheet

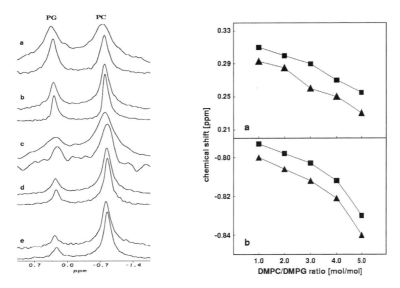

Figure 2. Left: ³¹P MAS NMR spectra (2kHz spinning speed, RT) of DMPC/DMPG vesicles before (top trace of each panel) and upon addition of Aβ₁₋₄₀ peptide at 30.1 L/P ratio (bottom trace). DMPC/DMPG molar ratios: 1:1 a); 2:1 b); 3:1 c); 4:1 d); 5:1 e). Right: Isotropic chemical shift values at RT for DMPG (a) and DMPC (b) as a function of DMPC/DMPG molar ratios of the membrane before (■) and upon addition of peptide (▲).

aggregates and bilayer surface charge density. In contrast, Aβ₁₋₄₀ incorporated into liposomes of the same composition shows an opposite behaviour.

<u>³¹P MAS NMR:</u> Since CD-spectroscopy does not provide a detailed view at a molecular level of the interactions of the peptide with membrane surfaces, ³¹P MAS NMR was used to study the nature and specificity of interactions of Aβ with the various lipid components when bound to charged membrane surfaces [24]. Samples were prepared as for CD-spectroscopy except for sonication, and the corresponding NMR lineshapes are shown in Figure 2 together with the spectra obtained for pure vesicles of different PC/PG content. Two resonances corresponding to DMPG and DMPC lipids can clearly be resolved with the intensity ratio changing from 1:1 to 5:1 PC/PG composition as expected. Upon addition of Aβ peptide, no specific interactions between the peptide and a single lipid component were observed, both resonances were effected in the same way. Line narrowing occurs for both resonances, most likely reflecting an increased fluidity of the system. A close inspection of the isotropic chemical shift values shows two detectable effects as summarized in Figure 2 (right) where the chemical shifts are plotted against the lipid composition before and upon binding of peptide. First, the chemical shift values change for both lipids depending on the content of charged lipids. Secondly, upon binding, Aβ induces a change in the chemical shift values for both

Extracellular Domain

     1   14

β -hairpin NH_3^+-Asp-Ala-Glu-Phe-Arg-His-Asp-Ser-Gly-Tyr-Glu-Val-His-His-

           15   27

middle helix   Gln-Lys-Leu-Val-Phe-Phe-Ala-Glu-Asp-Val-Gly-Ser-Asn-

Transmembrane Domain

         28   40

C-term helix   Lys-Gly-Ala-Ile-Ile-Gly-Leu-Met-Val-Gly-Gly-Val-Val-COO-

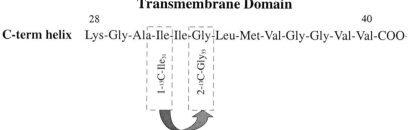

Figure 3. Amino acid sequence of amyloid-β peptide. The three domains of the monomeric membrane-bound secondary structure of Aβ₁₋₄₀ are indicated as predicted from modelling approaches [24]. Synthetically introduced residues carrying ^{13}C labels as used for rotational resonance distance measurements are marked.

resonances in the same direction, identical to the one observed when lowering the amount of charged vesicles. This effect reflects a partial compensation of membrane surface charge and suggests a mainly electrostatic binding of Aβ to the membrane surface.

The CD and ^{31}P MAS NMR studies clearly show that precise control over the lipid-peptide interactions and related parameters is essential for extracting high resolution structural information from this complex, disordered system where the conversion from a monomeric form into toxic aggregates has serious neurotoxicological implications.

^{13}C CP RR MAS NMR: To explore the potential of RR NMR for obtaining structural data for membrane-bound Aβ, the peptide was incorporated in a predominantly α-helical form into membranes in order to to determine the secondary structure of the transmembrane part of the peptide before major aggregation occurs. ^{13}C CP RR MAS NMR experiments were carried out on Aβ₁₋₄₀ which was specifically labelled as indicated in Figure 3 and reconstituted into DMPC-d_{67} bilayers at a 20:1 lipid/peptide molar ratio by cosolubilization.

In Figure 4 ^{13}C CP MAS NMR spectra are shown for labelled Aβ₁₋₄₀ peptide before (100.6 MHz) and after incorporation into membranes (125.7 MHz), respectively. In the spectrum for the solid peptide at 293 K (Trace a) the resonance at 175 ppm can be assigned to the labelled 1-^{13}C-Ile₃₁ position and the resonance at 44 ppm to the 2-^{13}C-Gly₃₃ position, both situated in the transmembrane part of the peptide. The spectrum of

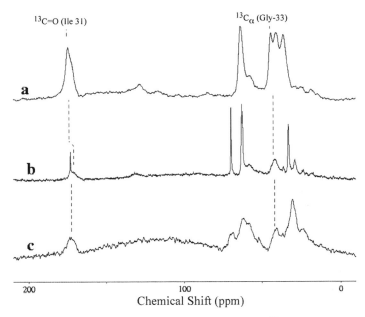

Figure 4. ^{13}C-*CP-MAS NMR spectra of* $A\beta_{1-40}$, *isotopically* ^{13}C *labelled as indicated: a) Spectrum (100.6 MHz) obtained at 9 kHz spinning speed and 293 K for solid* $A\beta_{1-40}$; *b) spectrum obtained at 125.7 MHz frequency for* $A\beta_{1-40}$ *incorporated into DMPC-d_{67} lipid bilayers at 293 K and 5 kHz spinning speed; c) spectrum as in b) but at 213 K and 8 kHz spinning rate.*

labelled Aβ upon incorporation into DMPC bilayers obtained at RT and 5 kHz spinning speed is shown in Trace b) of Figure 4. It is immediately obvious that the use of perdeuterated DMPC reduces the natural abundance signals arising from lipid carbon atoms drastically due to the missing CP conditions and reveals otherwise hidden resonances from the peptide and a few signals from the lipid glycerol backbone carrying protons [19,20]. In the membrane, the resonance of the carbonyl group of Ile31 in the transmembrane region is shifted upfield to 172 ppm and the 13Cα resonance of Gly33 up to 42 ppm at 293K. To observe peptide resonances at this temperature in a liquid-crystalline membrane is very surprising since, in most cases, the interference between the molecular motion of peptides or proteins and the coherent manipulation of nuclear magnetization causes a reduction of the ^{13}C-NMR signal intensity under CP-MAS conditions [25]. Therefore, it is usual practice to freeze out these motions in order to gain signal intensity [19]. The restricted dynamics for Aβ are likely due to interactions with other peptide molecules and/or an anchoring effect of the extracullar peptide part in agreement with similar observations obtained for other transmembrane systems like melittin or the M13 coat protein in liquid-crystalline membranes [20,27,28].

At 213K (see Trace c), the resonance positions from both labelled residues remain unchanged but overlap partially with lipid resonances which are now also broadened due to restricted motional mobility [20]. Comparison of the spectrum at 293 K with the one obtained at 213 K shows the same inhomgeneous linebroadening for both resonances arising from the peptide when incorporated into membranes. Therefore for this peptide segment, a distribution of conformational states must exist.

Despite the already restricted dynamics of the peptide at RT, rotational resonance experiments on the multilamellar DMPC lipid vesicles containing Aβ peptide were carried out at 213 K to exclude any dynamical effect which could interfere with the correct measurement of dipolar couplings and hence distances. At 213 K the peptide is also prevented from undergoing secondary structural changes into β-sheet aggregation. Magnetization exchange experiments under rotational resonance n=2 conditions were carried out for the membraneous system at 125.7 MHz ^{13}C frequency and 8172 Hz spinning speed between the 1-^{13}C-Ile$_{31}$ and 2-^{13}C-Gly$_{33}$ resonances. For off-resonance experiments a spinning speed of 7171 Hz was used. To analyse the magnetization exchange correctly, the amount of natural abundance signal arising from non labelled residues of the peptide was determined. Experiments identical to the one shown in Figure 4a were carried out on unlabelled solid Aβ$_{1-40}$ (spectra not shown) and compared to the one obtained for labelled peptide. For the carbonyl resonance, the percentage of natural abundance was found to be 33% and for the C$_{\alpha}$ region a fraction of around 30% was estimated.

The magnetization exchange curves obtained for the labelled Aβ in DMPC membranes at 213 K and n=2 conditions are shown in Figure 5. A signal decay can be seen indicating a medium distance between both labels in the transmembrane part of the peptide. The error limits (< 30%) for the peptide in the membrane is quite large but mainly caused by the existing large distribution of conformational states and the large natural abundance background. Unfortunately, RR conditions are not fulfilled for all conformational states at the same spinning speed due to the large inhomogeneous linebroadening, a problem already discussed before [20]. All these factors make a proper measurement of magnetization exchange and precise determination of an internuclear distance difficult, as seen in the error margin and fluctuating magnetization exchange values. Simulations of the magnetization decay curve still provides a lower and upper limit for the distance constraints [19-22]. However, due to the limited number of time points and the errors in the individual intensities, only a rough estimation of the internuclear distance can be done. Nevertheless the signal decay as seen in Figure 5 indicates α-helical structural features for the transmembrane segment of Aβ in agreement with the CD-measurements (s. Figure 1d), and similar to measurements obtained for the transmembrane helical part of M13 coat protein in membranes [20]. Any major β-sheet structure can be excluded based on the measured distances and the relevant CD spectrum. This result is not surprising since the method of incorporation relies on a water-free technique using the membrane mimicking solvent TFE. Due to the cosolubilization with lipids, the peptide is already in a stable lipid environment before

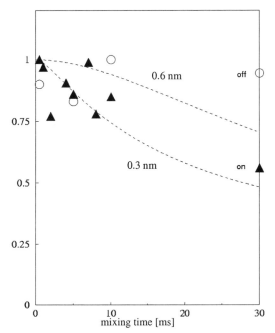

Figure 5: Magnetization exchange curves between 1-^{13}C-Ile$_{31}$ and 2-^{13}C-Gly$_{33}$ in Aβ$_{1-40}$ peptide incorporated into DMPC-d$_{67}$ membranes. Mixing times using a standard DANTE sequence for selective inversion [20] were varied between 0.5ms and 30ms. The difference magnetization of the spin pair $<I_z-S_z>$ is plotted over the mixing time. Experimental error indicated in text. Measurements were carried out at 125.7 MHz ^{13}C resonance frequency at 213K and n=2 RR conditions. Simulations are displayed for the upper and lower limit.

resuspension in an aqueous medium and, through the lipid matrix, protected against any conformational changes into β-sheet like structures when kept at low temperatures.

In this report first data are presented on the structure of the membrane anchored part of the Aβ$_{1-40}$ peptide when embedded in lipid bilayers. A combination of CD-spectroscopy and rotational resonance NMR using specifically ^{13}C labelled Aβ-peptide revealed that its transmembrane part exhibits a mainly α-helical secondary structure. But the experiments clearly show the problems of acquiring distance constraints for non-crystalline, disordered and heterogeneous membrane systems by means of solid state NMR.

Limits and Future Prospectives

For a precise determination of the secondary structure for Aβ-peptide in membranes and changes occuring upon onset of aggregation, the limits of RR and standard solid state

NMR approaches are severe, and new NMR strategies have to be used to obtain the required information for non-crystalline, disordered biological polymers. These strategies should also include the application of multiple or uniformly labelled systems to extract a huge wealth of structural information and avoid the expensive and difficult synthesis of specifically labelled molecules to obtain a single distance or torsion angle.

To obtain information about the modulation of the structural and aggregational properties of Aβ-peptide, various problems have to be addressed. Many problems are avoided in current solid state NMR studies by using peptides in a well defined crystalline environment to obtain good resolution. However, this approach not only avoids the problems related to the presence of many conformational states, but often also any relevance to the biological situation. Since the aggregation process in nature is not directed towards an highly ordered system but like in many other similar diseases into disordered and aggregated systems, NMR methodology has to adapt to this situation.

As seen in the ^{13}C MAS NMR spectra presented here, two sets of problems have to be addressed. First, the high natural abundance occurring from the lipid matrix (perdeuterated lipids are not always available) and other peptide residues makes accurate identification and intensity measurements of labeled residues difficult. Secondly, non-crystalline solids or peptides associated with biological membranes have inhomogeneously broadened NMR resonances because of structural heterogeneity present already in the non-aggregated state [20,29]. This puts serious limitations on the application of exact rotational resonance conditions and on the proper assignment for uniformly labeled systems.

One way to remove background signals due to natural abundance is the application of double quantum NMR filters, as demonstrated successfully up to medium distances (< 4Å) e.g. by Levitt [30], or Gregory et. al. using a DQ-DRAWS sequence [31]. But the efficiency is extremely low for longer distances (10% for 3.8 Å [31]) and would be even less for the Aβ-peptide in this study (distance for α-helix: 4,4 Å). And the labeled peptide is still diluted in an excess of lipids and water. Therefore the use of multiple or uniformly labeled peptide by means of molecular biology is more promising for the future. In this way shorter distances e. g. between ^{13}C dipoles are present providing much higher efficiencies for these double-quantum filters [21]. However due to the multiple strong couplings increased linewidths occur and the observation of weak couplings is more difficult.

The second big advantage of multiple labeled systems is to extract many structural constraints simultaneously, important to describe the structure of larger systems. However, this requires a full assignment of the occurring resonances despite the increased linewidths. Various groups have already successfully developed appropriate NMR sequences for full assignment for multiple labelled peptides and proteins e.g. on crystalline SH3 [32], the lyophilized ubiquitin [33], or the chemotactic tripeptide using 2D and 3D ^{15}N-^{13}C-^{13}C chemical shift correlation NMR [34]. Recently, the group of de Groot demonstrated on large chlorosmal antennae complexes a successful assignment for a system with considerable structural heterogeneity using high-field 2D and 3D dipolar correlation methods [35]. These ongoing developments should enable researchers in the near future by using high-fields and 3D experiments to assign multiple labelled peptides and proteins in their disordered, and in case of Aβ also neurotoxic states, despite the presence of large inhomogenous linebroading.

Another unsolved problem is how to extract many distance and torsion angle constraints from multiple or uniformly labelled molecules with their inherent multiple strong couplings. Various broadband recoupling techniques for homonuclear and heteronuclear systems for distance measurements have been developed over the last years e.g. RFDR (rf-driven recoupling), DRAMA (dipolar recovery at the magic angle), SEDRA (simple excitation for the dephasing of rotational-echo amplitudes), C7, Post-C7 [21]. However, the conversion of cross peak intensities into distance information (similar to NOE constraints in solution NMR) is difficult and therefore the extraction of multiple distance information from these spectra is currently a huge challenge. Similar problems arise for the determination of torsion angles. Recently, Hong and coworkers developed a promising approach which could be suitable for samples with a broad distribution of various secondary structural features like Aβ [36]. Their NMR pulse sequences discriminate between α-helix and β-sheet residues and filter them selectively. In this way it should be possible to determine the amount of α-helix and β-sheet residues in a molecule at various states. In addition, development is also ongoing how to determine the number of molecules arranged into aggregates using static multiple quantum NMR techniques [17].

To obtain a complete structural description of Aβ-peptide associated with membranes at its various aggregational states, MAS NMR techniques have to be applied to multiple/uniformly labeled peptide in the future. This requires development in both spectral resolution and sensitivity, and includes high field NMR machines, development of labeling schemes and sample preparation for improved resolution and finally the development of pulse sequences and appropriate algorithms to extract multiple distance and torsion angle constraints from these systems.

Acknowledgements

Financial support from BBSRC (43/B11683), EU (FMRX-CT96-0004) and local funding (Umeå University) is acknowledged. CG is recipient of a DFG Emmy-Noether Fellowship and A. W. of a BBSRC Senior Fellowship (43/SF09211).

References

[1]   Masters, C. L., Simms, G., Weinman, N. A., Multhaup, G., McDonald, B. L., and Beyreuther, K., Proc. Nat. Sci. USA 82 (1985) 4245.
[2]   Haass, C., and Selkoe, D. J., Cell 75 (1993) 1039.
[3]   Rochet, J.-C., and Lansbury Jr., P. T., Curr. Opin. Struct. Biol. 10 (2000) 60.
[4]   Lansbury Jr., P. T., Proc. Natl. Acad. Sci. USA 96 (1999) 3342.
[5]   Burkoth, T. S., Benzinger, T. L. S., Urban, V., Morgan, D. M., Gregory, D. M., Thiyagarayan, P., Botto, R. E., Meredith, S. C. and Lynn, D. G., J. Am. Chem. Soc. 122

(2000) 7883.

[6]   Iversen, L. L., Mortishire-Smith, R. J., Pollack, S. J. and Shearman, M. S., Biochem. J. 311 (1995) 1.

[7]   Lorenzo, A., and Yankner, B. A., Proc. Natl. Acad. Sci. USA 91 (1994) 12243.

[8]   Walsh, D. M., Hartley, D. M., Kusumoto, Y., Fezoui, Y., Condron, M. M., Lomakin, A. Benedek, G. B., Selkoe, D. J., and Teplow, D. B., J. Biol. Chem. 274 (1999) 25945.

[9]   Vargas, J., Alarcón, J. M. and Rojas, E., Biophys. J. 79 (2000) 934.

[10]   Terzi, E., Hölzemann, G., and Seelig, J., Biochemistry 36 (1997) 14845.

[11]   Kremer, J. J., Pallitto, M. M., Sklansky, D. J., and Murphy, R. M., Biochemistry 39 (2000) 10309.

[12]   McLaurin, J., Franklin, T., Chakrabartty, A., and Fraser, P. E., J. Mol. Biol. 278 (1998) 183.

[13]   Choo-Smith, L.-P., Garzon-Rodriguez, W., Glabe, C. G., and Surewicz, W. K., J. Biol. Chem. 272 (1997) 22987.

[14]   Kawahara, M., Arispe, N., Kuroda, Y., and Rojas, E., Biophys. J. 73 (1997) 9412.

[15]   Rhee, S. K., Quist, A. P., and Lal, R., J. Biol. Chem. 29 (1998), 13379.

[16]   Jarvet, J., Damberg, P., Bodell, K., Eriksson, L. E. G. , and Gräslund, A., J. Am. Chem. Soc. 122 (2000) 4261.

[17]   Antzutkin, O. N., Balbach, J. J., Leapman, R. D., Rizzo, N. W., Reed J., and Tycko, R., Proc. Natl. Acad. Sci. USA 84 (2000) 13045.

[18]   Mason, R. P., Jacob, R. F., Walter, M. F., Mason, P. E., Avdulov, N. A., Chochina, S. V., Igbavboa, U., and Wood, W. G., J. Biol. Chem. 274 (1999) 18801.

[19]   Smith, S. O., Aschheim, K. and Groesbeek, M., Quart. Rev. Biophysics 29 (1996) 395.

[20]   Glaubitz, C., Gröbner, G., and Watts, A., Biochim. Biophys. Acta 1463 (2000) 151.

[21]   Griffin, R. G., Nature Struct. Biol. 5 (1998) 508.

[22]   Raleigh, D. P., Levitt, M. H., and Griffin, R. G., Chem. Phys. Lett. 146 (1988) 71.

[23]   Andrade, M. A., Chacon, P., Merelo, J. J., and Moran, F., Protein Engineering 6 (1993) 383.

[24]   Pinheiro, T. J. T., and Watts, A., Biochemistry 33 (1994) 2459.

[25]   Warschawski D. E., Gross, J. D., and Griffin, R. G., J. Chim. Phys. PCB 95 (1998) 460.

[26]   Durell, S. R., Guy, H. R., Arispe, N., Rojas, E., and Pollard, H. B., Biophys. J. 67 (1994) 2137.

[27]   McDonnell, P. A., Shon, L., Kim, Y., and Opella, S. J., J. Mol. Biol. 233 (1993) 447.

[28]   Naito, A., Nagao, T., Norisada, K., Mizuno, T., Tuzi, S., and Saitô, H., Biophys. J. 78 (2000) 2405.

[29]   Tycko, R., J. Biomol. NMR 8 (1996) 239.

[30]   Karlsson, T., Edén, M., Luthman, H., and Levitt, M. H., J. Magn. Reson. 145 (2000) 95.

[31]   Gregory, D. M., Wolfe, G. M., Jarvie, T. P., Sheils J. C., and Drobny, G. P., Mol. Phys. 89 (1996) 1835.

[32]   Pauli, J., van Rossum, B., Förster, H., de Groot, H. J. M., and Oschkinat, H., J. Magn. Reson. 143 (2000) 411.

[33]   Hong, M., J. Biomol. NMR 15 (1999) 1.

[34]   Rienstra, C. M., Hohwy, M., Hong, M., and Griffin, R. G., J. Am. Chem. Soc. 122 (2000) 10979.

[35]   van Rossum, B.-J., Steensgaard, D. B., Mulder, F. M., Boender, G. J., Schaffner, K., Holzwarth, A. R., and de Groot, H. J. M., Biochemistry, in press.

[36]   Huster, D. Yamaguchi, S., and Hong, M., J.Am. Chem. Soc. 122 (2000) 11320.

Photochemically induced dynamic nuclear polarization in bacterial photosynthetic reaction centres observed by ^{13}C solid-state NMR

Jörg Matysik,[a] Alia,[a,b] Peter Gast,[b] Johan Lugtenburg,[a] Arnold J. Hoff,[b] Huub J. M. de Groot[a]

[a]*Leiden Institute of Chemistry, Gorlaeus Laboratoria, p.o. box 9502, and*
[b]*Department of Biophysics, Huygens Laboratorium, p.o. box 9504,*
2300 RA Leiden, The Netherlands

Introduction

Photosynthesis, the synthesis of organic compounds upon utilization of light energy, is one of the key reactions for life on earth. The first step of this complex process, a light induced electron transfer, occurs in photosynthetic reaction centre (RC) membrane proteins. The electron is emitted from a (bacterio)chlorophyll (BChl) aggregate in its electronically excited state. In bacterial RCs, this aggregate is a strongly coupled BChl *a* dimer, the so-called "special pair" (P). When photoexcited to a higher electronic singlet state, P transfers an electron to the primary acceptor, a pheophytin molecule, within 3 ps. The electron is then transferred to Q_A in about 200 ps. Subsequently, in a much slower reaction taking about 100 μs, the electron is transferred to the ultimate electron acceptor Q_B. This light-induced electron transfer sequence is repeated after the special pair has been re-reduced by a cytochrome. Although the spatial structure and kinetics of several RCs are known to atomic resolution, there is no clear understanding of the process of electron emission from the electronically excited primary electron donor. In addition, a detailed picture of the molecular mechanism of the inhibition of the back reaction, which is probably due to the high exothermic reaction enthalpy pushing the system into the inverted Marcus region, is missing. The functionally crucial electronic structure can be probed by spectroscopic methods. Grosso modo, vibrational spectroscopy provides information about electron densities *between* nuclei, and NMR spectroscopy *at* the nuclei.

S.R.Kiihne and H.J.M.deGroot (eds.), Perspectives on Solid State NMR in Biology, 215-225.

Photochemically induced dynamic nuclear polarization (photo-CIDNP) is a method to enhance the intensity of an NMR line by inducing a non-Boltzmann distribution of the nuclear spin states. Photo-CIDNP is well-known from liquid NMR [1], and a reaction mechanism for photo-CIDNP has been described in terms of a radical-pair mechanism [2,3]. Photo-CIDNP can also be observed by solid-state NMR spectroscopy. It has been detected in frozen samples of bacterial reaction centres by Zysmilich and McDermott [4,5,6] and in plant reaction centres by us [7]. The discovery of the magnetic field effect [8,9], which is the dependence of the triplet quantum yield on the strength of the external magnetic field, opened the view on the amazing magnetic properties of photosynthetic reaction centres. Since this effect has been interpreted in terms of nuclear couplings on the mixing rate of the radical pair, the possibility of photo-CIDNP in reaction centres was already predicted at an early stage [10]. Due to an electron-electron-nuclear three-spin mixing mechanism, the spin-correlated radical pair polarises nuclei with Zeeman frequencies close to a matching condition that corresponds to the difference of the Zeeman energies of the two electrons [11,12]. Nuclear coherences generated by spin-correlated radical pairs have been observed by time-resolved EPR spectroscopy [13,14]. Sorting of nuclear spins can occur by very fast recombination of the radical pair since the molecular triplet has left the three-spin system [15]. Therefore, the spatial pattern of the nuclear polarization reflects the spin space of the spin-correlated radical pair. Additional polarization can be obtained by the different nuclear relaxation kinetics of the singlet and paramagnetic triplet species [15]. Therefore, a high nuclear polarization can be achieved, providing a unique opportunity to study the primary events of photosynthesis on the atomic scale.

In this paper we describe the illumination set-up we designed for a standard wide bore MAS NMR probe. We report our recent photo-CIDNP data observed on photosynthetic RCs from wild-type and a carotenoidless mutant (R26) of *Rhodobacter sphaeroides*.

Methods

Sample preparation. The RCs from Rb. sphaeroides R-26 were isolated by the procedure of Feher and Okamura [16]. The RCs from wild-type Rb. sphaeroides were isolated using the method of Feher and Okamura with slight modifications. The removal of Q_A was done by incubating the reaction centres at a concentration of 0.6 µM in 4% LDAO, 10 mM o-phenanthroline, 10 mM Tris buffer, pH 8.0 for 6 hrs at 26 °C, followed by washing on a DEAE column and removal of the reaction centres from the column with 0.5 M NaCl in 10mM Tris buffer pH 8.0 containing 0.025% LDAO and 1mM EDTA [17]. Quinone reduction in the RC was performed by addition of 75 mM sodium ascorbate followed by freezing under illumination in the NMR probe in the magnet. Approximately 5 mg of the RC protein complex embedded in LDAO micelles was used for NMR measurements.

MAS-NMR measurements. The NMR experiments have been performed using MSL-400 and DMX-400 NMR spectrometers (Bruker GmbH, Karlsruhe, Germany) equipped with a double-resonance magic angle spinning (MAS) probe working at 396.5 MHz for ^{1}H and 99.7 MHz for ^{13}C. The sample was loaded into a 4-mm clear sapphire rotor and inserted into the MAS probe. ^{13}C MAS NMR spectra were obtained with a spinning frequency $v_r = 4$ kHz at a temperature of 225 K. At the start of the experiments, the sample was frozen slowly with liquid nitrogen-cooled bearing gas, using slow spinning of $v_r = 600$ Hz to ensure a homogeneous sample distribution against the rotor wall [18]. To obtain spectra under illumination, the sample was continuously irradiated from the side of the spinning sapphire rotor. The modifications of the probe and the entire illumination setup constructed for a standard Bruker wide bore MAS NMR spectrometer is described in the results and discussion section. The light and dark spectra have been collected with a Hahn echo pulse sequence and TPPM proton decoupling [19]. Typically, a recycle delay of 15 s was used, and a total of 24.000 scans per spectrum were collected over a period of 24 h.

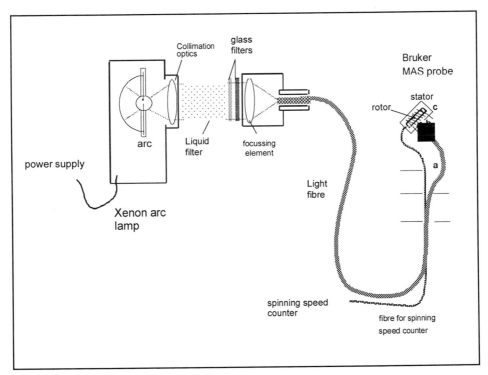

Fig. 1. Schematic diagram of the illumination system. The letters a, b, c represent the places in the probe where modifications were made.

Results and Discussion

The MAS illumination set-up.
The light illumination set-up comprises a 1000 Watt Xenon arc lamp with collimation optics (model 66021, Oriel, Stratford, CT, USA), a liquid filter and glass filters, a focusing element and a light fiber (Fig. 1). Since the emission spectrum of a Xe lamp is similar to sunlight, the full range of radiation from UV to IR was available for illumination. Any interference of the incident radiation with the spinning speed counter, working in the near-IR region, was avoided by cutting off the near-IR wing with the glass filters. A fiber bundle was used to transport the light from the collimation optics into the probe, which was specially constructed for this purpose. The light fiber provides high optical transparency in a broad spectral range, as well as mechanical flexibility for being fixed onto the stator of the MAS probe. The coating of the fiber does not contain any metal to avoid arcing inside the probe. Some modification of the MAS probe was necessary to approach the sample from the side. This

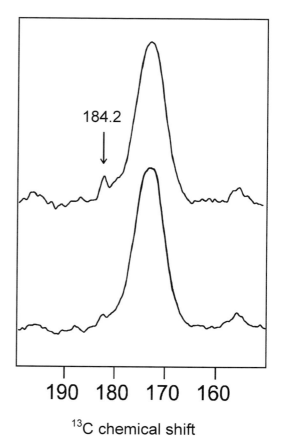

Fig. 2. ^{13}C MAS NMR spectrum of a bacterial reaction centre. The intensity of the C-4 signal of Q_A at 184.2 ppm in dark (A) is reduced upon irradiation of the sample during the NMR measurement (B).

includes (a) an opening drilled into the upper most metal grounding plate separating the stator chamber from the RF electronic parts of the probe, (b) drilling a small opening into the stator, and (c) winding a new coil from thin round silver wire of 1.0 mm diameter. The inner diameter of the coil (4.2 mm) is close to the outside diameter of the rotor (4.0 mm). With the custom-made coil, the 90^0-carbon pulse length of 5 μs is obtained with an RF power of ~250 W. For 90^0-proton pulses, a pulse length of 4 μs is obtained at ~75 W.

The excitation intensity within the MAS probe. To determine if the number of photons transported via the fiber into the stator is sufficient to penetrate the sample completely, the following method has been applied. A sample of bacterial reaction centres containing a quinone Q_A which is ^{13}C-isotope labelled at the C-4 position [20], is measured with cross-polarization in the dark at 220 K (Fig. 2A). Under illumination with white light, the quinone radical anion $Q_A^{-\bullet}$ is formed and the signal of the ^{13}C isotope labelled quinone is reduced by about 70% (Fig. 2B). Due to the absence of Q_B, the lifetime of the charge separated $P^{+\bullet} Q_A^{-\bullet}$ pair is around 50 ms [21]. If we ignore the high optical density of the sample at several wavelengths and assume homogenous illumination throughout the sample, an average value of about 50 photons per second per RC is estimated for the light excitation intensity. This compares well with the 200 photons per second per RC obtained from a simulation of the light saturation of the photo-CIDNP signal [15]. Because of the high optical density of the solid-state NMR sample, the true light intensity used in our experiments is probably higher than 50 photons per second per RC. Thus the excitation intensity may be near to the saturation level for photo-CIDNP for the primary radical pair.

Photo-CIDNP in RCs from carotenoidless mutants (R26) Figure 3 shows the ^{13}C MAS NMR spectra of a ^{13}C natural abundance sample of quinone-depleted reaction centres of *Rb. sphaeroides* R26, acquired at 220 K in the dark (A) and with continuous illumination using white light (B). The ^{13}C MAS NMR spectrum of the illuminated sample shows strong nuclear spin polarization in NMR lines compared to the dark. Both enhanced-absorptive (positive) and emissive (negative) lines appear in the ^{13}C NMR spectrum. The carbons of the aromatic ring gain special photo-CIDNP intensity enhancement. The emissive signals (110.5, 106.8, 101.5, 97.8 and 95.8 ppm) arise in the region of methine carbons. The observed chemical shifts match well with a neutral and non-radical state of a BChl *a*. A radical species may show paramagnetically shifted NMR lines and furthermore, the decay of the transient radical cation species is too fast for it to be observed on the NMR time scale. Therefore, the observed NMR spectrum shows the chemical shifts of neutral non-radical cofactors that are in the electronic ground-state but have strongly spin-polarized nuclei. Our data are similar to the results obtained for dense pellets of the same biological system [5]. Looking at the earlier data in the literature, we conclude that photo-CIDNP can be induced with better signal-to-noise in dilute frozen RC protein complex embedded in detergent micelles.

Experiments with various MAS spinning frequencies help to distinguish centrebands from sidebands (Figs. 3C & D). An iterative Herzfeld-Berger analysis [22,23] of the sideband intensities and symmetry indicate values of $\delta \approx 10$ kHz with $\eta \approx 1$ for the chemical shift anisotropy, which are typical for aromatic carbon atoms [24,25]. Almost all centrebands have been reconciled with a response from a single BChl *a* [5]. Table 1 summarises the assignments obtained in the earlier work and a tentative assignment for ^{13}C signals in comparison with the chemical shift data available for BChl *a* (see Fig. 3E for numbering). It is obvious that mainly the carbon atoms of the aromatic ring are enhanced. All emissive signals are assigned to methine bridges. It is

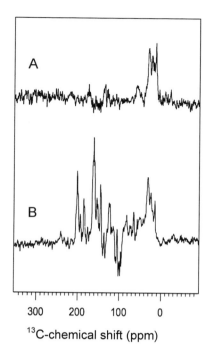

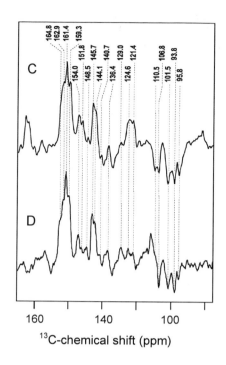

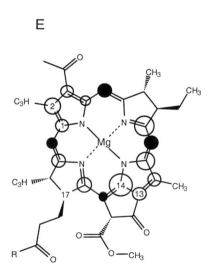

E

Fig. 3. ^{13}C MAS NMR spectra of a quinone depleted R26 bacterial reaction centre sample. In the dark, the aliphatic region can be detected (A), while strong photo-CIDNP signals appear upon continuous illumination with white light (B). The spinning frequency is 4 kHz, the temperature is 220 K. The low-frequency region of the ^{13}C photo-CIDNP MAS NMR spectra of the bacterial RC at spinning frequencies of 4 kHz (C) and 5 kHz (D). The centrebands are marked. The structure of a bacteriochlorophyll a molecule (E) with numbering according to the IUPAC nomenclature. The circles indicate atoms with high photo-CIDNP according to the tentative assignments. Open circles indicate positive signals, the filled circles indicates negative CIDNP intensity enhancement. For the tentative assignments, see Table 1.

Table 1: Tentative assignments of the ^{13}C photo-CIDNP NMR signals

BChl a		assignment	photo-CIDNP	
$\sigma_{liq}{}^a$	$\sigma_{ss}{}^a$	carbon no.	σ^b	σ^c
199.3	194.5	3^1	-	-
189.0	188.2	13^1	-	-
173.4	174.0	17^3	-	-
171.6	171.4	13^3	-	-
168.9	170.2	6	167.5 A	-
167.3	168.9	19	165.2 A	164.8 A
-	-	-	163.5 A	-
-	-	?	162.7 A	162.9 A
160.8	160.7	14	161.2 A	161.4 A
158.5	158.0	9	158.9 A	159.3 A
152.2	150.1	16	154.3 A	154.0 A
151.2	153.5	1	151.6 A	151.8 A
150.2	152.2	4	148.8 A	148.5 A
149.5	147.2	11	145.8 A	145.7 A
-	-	?	136.9 A	140.7 A
142.1	140.7	2	144.3 A	144.1 A
137.7	136.1	3	135.8 A	136.4 A
130.5	124.1	13	130.9 A	129.0 A
123.9	119.9	12	124.6 A	124.6 A
-	-	?	-	121.4 A
-	-	?	-	110.5 E
109.7	105.8	15	106.5 E	106.8 E
102.4	100.0	10	101.8 E	101.5 E
99.6	98.8	5	97.8 E	97.8 E
96.3	93.7	20	95.5 E	95.8 E

A = absorptive, E = emissive
[a] Egorova-Zachernyuk et al., to be published. The liquid NMR chemical shift data σ_{liq} have been obtained in acetone-d_6.
[b] Ref. [5]
[c] This work. Chemical shift value is averaged between experiments at different spinning speed.

remarkable that the emissive signals have very similar intensities. This indicates a rather symmetric electron spin density distribution within the macrocycle in the radical cation state, and is clearly distinguished from the highly asymmetric electron spin density pattern observed in photosystem II [7].

There are four additional signals, which cannot be assigned to a single BChl a molecule (162.9, 140.7, 121.4, and 110.5 ppm). The signal at 162.9 ppm occurs in the region of the strongest absorptive signals. Another unassigned signal, an emissive signal at 110.5 ppm, occurs rather close to another emissive signal. Also the two signals at 162.9 and 121.4 ppm may arise from another cofactor. The lineshape of several of the emissive signals appears asymmetric and the lines are rather broad. This observation suggests that the photo-CIDNP spectrum shows signals from more than one BChl *a* of the special pair, while most signals are overlapping and cannot be separated. Our most recent results obtained with samples containing ^{13}C-labelled cofactors provide evidence that both cofactors of the special pair as well as a BPheo are represented in the photo-CIDNP enhanced NMR spectra (Schulten et al., unpublished).

Photo-CIDNP in RCs from quinone depleted and quinone reduced samples of R26. Figure 4A shows the ^{13}C photo-CIDNP NMR spectra from quinone reduced RCs obtained from R26 and *Rhodobacter sphaeroides*. The quinone reduced RC from R26 gave similar photo-CIDNP spectra as obtained from Q-depleted RC (Fig. 3B). Slight differences can be related to the different intensities of the BPheo signals, caused by different kinetics at the acceptor site [6].

Photo-CIDNP in RCs from wild-type (WT) Rb. sphaeroides. Due to a much shorter triplet lifetime of 2 to 300 ns [26], experiments on WT are very interesting for studying the mechanism of photo-CIDNP. As depicted in Fig. 4B, photo-CIDNP was also observed in RCs from WT *Rb. sphaeroides*. The strength of the photo-CIDNP effect in R26 and WT samples is comparable. This means that the triplet state itself is not inducing the nuclear polarization as known from other solid-state systems [27]. Critical for the mechanism proposed by Polenova and McDermott [15] is a very fast decay of the primary triplet ($P\uparrow^+$ BPheo$\uparrow^-$) to the secondary triplet ($P\uparrow\uparrow$), which empties the primary triplet species in its nuclear spin sorting equilibrium with the singlet radical pair ($P\uparrow^+$ BPheo$\downarrow^-$). These events are not changed in the WT, therefore, the total nuclear polarization of R26 and WT samples should not be dramatically changed.

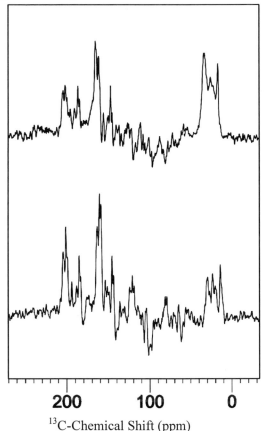

200   100   0

[13]C-Chemical Shift (ppm)

Fig. 4. ^{13}C photo-CIDNP MAS NMR spectra of quinone reduced bacterial reaction centres from WT (A) and R26 (B) from Rb. sphaeroides.

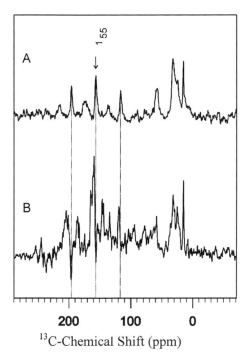

155

A

B

200   100   0

^{13}C-Chemical Shift (ppm)

Fig. 5. ^{13}C MAS NMR spectra of quinone reduced bacterial reaction centres from [4'-^{13}C]-tyrosine labelled Rb. sphaeroides (M)Y210W mutant in the dark (A) and under light (B).

There are some interesting differences between the photo-CIDNP spectra from RC of WT and from R26. These differences are more pronounced in the region of emissive signals. It provides evidence that the presence of the carotenoid effects the photo-CIDNP. In this case, the differences between the spectra of R26 and WT could in principle involve regions corresponding to the ^{13}C response of a carotenoid, i.e. between 120 and 130 ppm [28]. In that spectral range, however, no remarkable changes can be observed. On the other hand, it is possible that the different kinetics causes a different combination of the complex intensity pattern of overlapping absorptive and emissive lines of the various cofactors, as well as of the surrounding protein residue (see below) to an extent that the observed envelope is significantly modified. Further studies with labelled WT RC are underway to investigate these differences in more detail.

Photo-CIDNP observed in the apoprotein. To probe whether the carbon atoms of amino acid residues around the cofactors are spin-polarized, either directly by the electron spin or indirectly via nuclear Overhauser effect from other carbon atoms, we investigated an RC sample in which all tyrosine residues are ^{13}C labelled at the 4'-position [29]. The RC of this sample was mutated on the position (M)210, where a tyrosine is exchanged against a tryptophane. The tyrosine-(M)210 is very close to P, and has shown to be of importance to the electron transfer processes. In particular, its exchange to tryptophane slows down the electron transfer by a factor of four [29]. The remaining tyrosine residues are not directly involved in electron transfer. Finally, the sample contains a carotenoid. Figure 5A shows the MAS NMR spectrum of the dark sample. The ^{13}C-labelled nuclei are clearly visible at 155 ppm. Also a sideband pattern characteristic for aromatic atoms is observed. At around 175 ppm, the broad peptide carbonyl background signal with its sidebands is visible. Upon illumination (Fig. 5B), emissive lines of the 4'-^{13}C atoms of the tyrosine residues occur, overlapping the photo-CIDNP envelope of the unlabelled RC. It is at present not clear how many and which tyrosine residues are involved in photo-CIDNP. The observation of an emissive NMR signal at the position 4'-C is in line with liquid ^{1}H photo-CIDNP NMR which showed emissive lines of the 3'-H and 5'-H protons, while the other half of the ring (2'-H, 6'-H) including the CH$_2$ side group show enhanced absorptive signals [30]. The photo-CIDNP effect on the ^{13}C-labelled tyrosine is in the same order of magnitude as the natural abundance signals from the cofactors. These data contrast with the strong photo-CIDNP observed from an isotope labelled histidine coordinating to a BChl which showed an intensity comparable to the labelled cofactor atoms [6]. Currently, further photo-CIDNP studies with bacterial RC samples containing isotope labels in the apoprotein are in progress.

Conclusions

The results presented in this paper demonstrate that photo-CIDNP MAS NMR spectroscopy provides an excellent tool to study the electronic structure of photosynthetic reaction centres on the atomic level. This opens the way for extracting the basic mechanistic principles of electron transfer and inhibiting the back reaction of highly optimized natural electron pumps, which may be of help in guiding the construction of artificial photosynthetic devices.

Acknowledgments

The authors thank Dr. S. Shochat for providing RCs from tyrosine-(M)210 of *Rb. sphaeroides*, Dr. W.B.S. van Liemt and Dr. B. van Rossum for preparing the Q$_A$ labelled sample, and B.M.M. Joosten for the help in the purification of the RCs. The kind help of K. Erkelens, F. Lefeber and J.G. Hollander in the operation of the NMR spectrometers is acknowledged. Thanks to Els Schulten for making her results available prior to publication. J.M. acknowledges a Marie Curie fellowship of the European Commission (ERB4001 GT972589) and a Casimir-Ziegler award of the Academies of Sciences in Amsterdam and Düsseldorf. This work was financially supported by the PIONIER programme of the Netherlands Organization for Scientific Research (NWO).

References

[1]   Hore, P. J. and Broadhurst, R. W., Progr. NMR Spectrosc. 25 (1993) 345.
[2]   Kaptein, R., in Muss, L.T., Atkins, P.W., McLauchlan K.A. and Pederson J.B. (eds.), Introduction to Chemically Induced Magnetic Polarization, Reidel, Dordrecht, 1977, p. 257.

[3]   Müller, F., Schagen, C. G. V. and Kaptein, R., Meth. Enzymol. 66 (1980) 385.

[4]   Zysmilich, M. G. and McDermott, A.E., J. Am. Chem. Soc. 116 (1994) 8362.

[5]   Zysmilich, M. G. and McDermott, A.E., Proc. Natl. Acad. Sci. USA 93 (1996) 6857.

[6]   Zysmilich, M. G. and McDermott, A.E., J. Am. Chem. Soc. 118 (1996) 5867.

[7]   Matysik, J., Alia, Gast, P., van Gorkom, H.J., Hoff, A.J. and de Groot, H.J.M., Proc. Natl. Acad. Sci. USA 97 (2000) 9865.

[8]   Hoff, A.J., Rademaker, H., van Grondelle, R. and Duysens, L.N.M., Biochim. Biophys. Acta 460 (1977) 547-554.

[9]   Blankenship, R.E., Schaafsma, T.J. and Parson W.W., Biochim. Biophys. Acta (1977) 297-305.

[10]   Goldstein, R.A., Boxer, S.G., Biophys. J. 51 (1987) 937.

[11]   Jeschke, G., J. Chem. Phys. 106 (1997) 10072.

[12]   Jeschke, G., J. Am. Chem. Soc. 120 (1998) 4425.

[13]   Weber, S., Berthold, T., Ohmes, E., Thurnauer, M.C., Norris, J.C. and Kothe G., Appl. Magn. Reson. 11 (1996) 461.

[14]   Kothe, G., Bechthold, M., Link, G., Ohmes, E. and Weidner J.U., Chem. Phys. Let. 283 (1998) 51.

[15]   Polenova, T. and McDermott, A.E., J. Phys. Chem. B 103 (1999) 535.

[16]   Feher, D. and Okamura, M.Y., in Clayton, R.K. and Sistrom, W.R. (eds.) The Photosynthetic Bacteria, Plenum, New York, 1978, p. 349.

[17]   Okamura, M.Y., Isaacson, R.A. and Feher. G., Proc. Natl. Acad. Sci. U.S.A. 72 (1975) 3491.

[18]   Fischer, M.R., de Groot, H.J.M, Raap, J., Winkel, C., Hoff, A.J., and Lugtenburg, J., Biochemistry 31 (1992) 11038.

[19]   Bennet, A.E., Rienstra, C.M., Auger, M., Lakshmi, K.V. and Griffin, R.G., J. Chem. Phys. 103 (1995) 6951.

[20]   van Liemt, W.B.S., Boender, G.J., Gast, P., Hoff, A.J., Lugtenburg, J. and de Groot, H.J.M., Biochemistry 34 (1995) 10229.

[21]   Romijn, J.C. and Amesz, J., Biochim. Biophys. Acta 423 (1976) 164.

[22]   Herzfeld, J. and Berger, A.E., J. Chem. Phys. 73 (1980) 6021.

[23]   de Groot, H.J.M., Smith, S.O., Kolbert, A.C., Courtin, J.M.L., Winkel, C., Lugtenburg, J., Herzfeld, J. and Griffin, R.G., J. Magn. Reson. 91 (1991) 30.

[24]   Veeman, W.S., Progr. NMR Spectrosc. 16 (1984) 193.

[25]   Metz, G., Siebert, F. and Engelhardt, M., Biochemistry 31 (1992) 455.

[26]   Bosch, M.K., Gast, P., Franken, E.M., Zwanenburg, G., Hore, P.J. and Hoff, A.J., Biochim. Biophys. Acta 1276 (1996) 106.

[27]   van den Heuvel, D.J., Schmidt, J. and Wenckebach, W.Th., Chem. Phys. 187 (1994) 365.

[28]   de Groot, H.J.M., Raap, J., Winkel, C., Hoff, A.J. and Lugtenburg, J., Biochemistry 31 (1992) 12446.

[29]   Shochat, S., Gast, P., Hoff, A.J., Boender, G.J., van Leeuwen, S. and van Liemt, W.B.S., Vijgenboom, E., Raap, J., Lugtenburg, J. and de Groot H.J.M., Spectrochim. Acta 51A (1995) 135.

[30]   Kaptein, R., Dijkstra, K., Müller, F., Schagen, C.G.V. and Visser, A.J.W.G., J. Mag. Res. 31 (1978) 171.

INDEX